ROSELLE PUBLIC LIBRARY DISTRICT
3 3012 00006 6882

W9-AAH-246

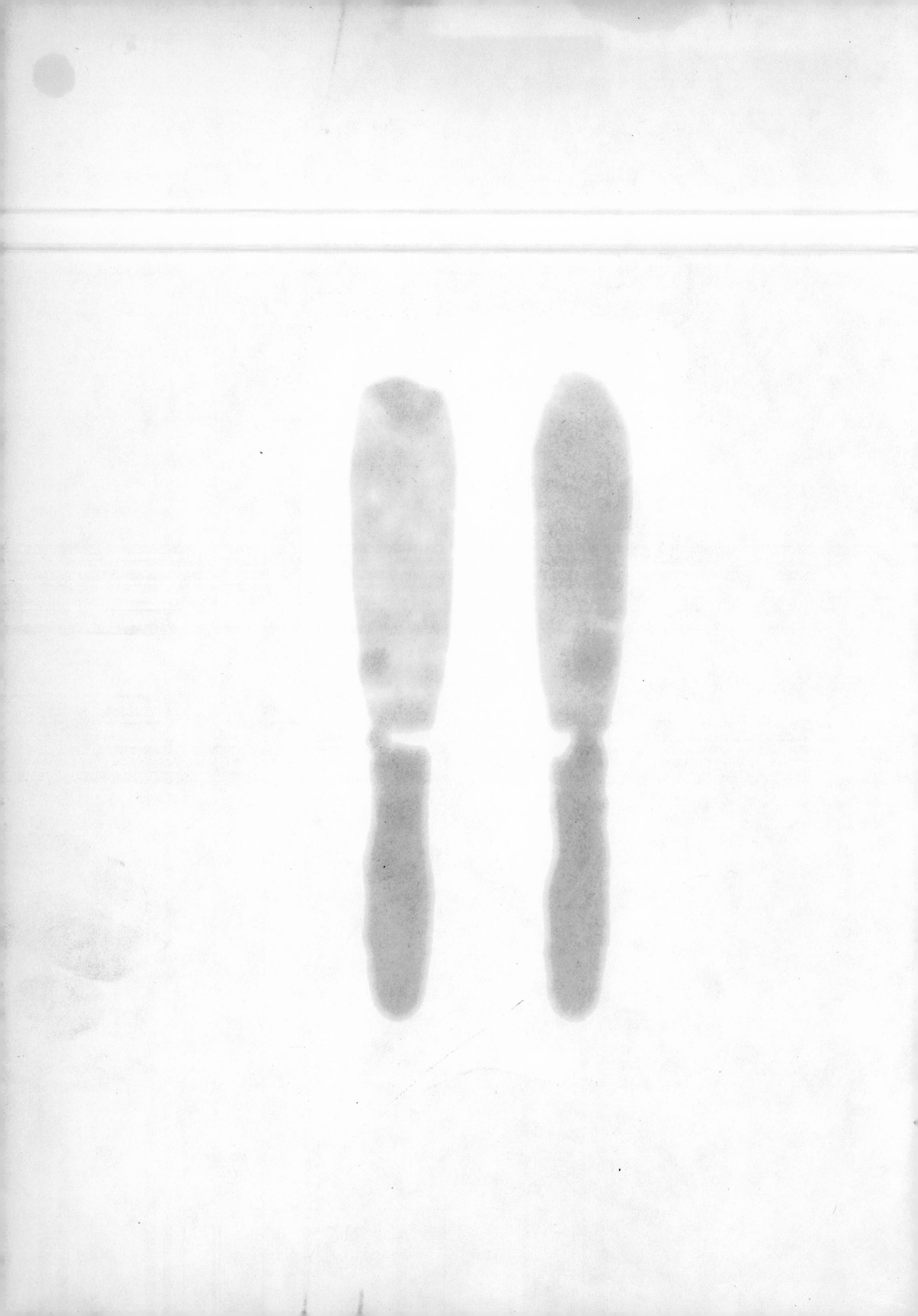

The Rand McNally Library of Astronomical Atlases
for Amateur and Professional Observers
Series Editors Garry Hunt and Patrick Moore

Saturn

Garry Hunt
and Patrick Moore

Foreword by Professor Archie E. Roy

Published in Association with
the Royal Astronomical Society

Rand McNally and Company
New York · Chicago · San Francisco

Saturn

ISBN 528-81545-8
Library of Congress Catalog Number 82-60035

Saturn was edited and designed by
Mitchell Beazley International Ltd,
Mill House, 87–89 Shaftesbury Avenue,
London W1V 7AD

Phototypeset by Servis Filmsetting Ltd
Origination by Adroit Photo Litho Ltd
Printed in the United States

Editor Judy Garlick
Designer Wolfgang Mezger
Editorial Assistant Charlotte Kennedy
Picture Research Millicent Trowbridge

Executive Editor Lawrence Clarke
Art Manager John Ridgeway
Production Manager Barry Baker

The units and notation used throughout this book are based on the Système International des unités (SI units), which is currently being introduced universally for scientific and educational purposes. There are seven "base" units in the system: the *meter* (m), the *kilogram* (kg), the *second* (s), the *ampere* (A), the *kelvin* (K), the *mole* (mol) and the *candelo* (cd). Other quantities are expressed in units derived from the base units; thus, for example, the unit of force the newton (N) is defined as the force required to give a mass of one kilogram an acceleration of one meter per second squared ($kg\,m\,s^{-2}$).

Some branches of science continue to adhere to a few of the older units, and in one case an editorial concession has had to be made to existing scientific usage: the S1 unit of magnetism, the tesla, has been dropped in favor of the more common unit, the gauss. One tesla is equal to 10,000 gauss.

For very large and very small numbers, "index notation" has been adopted, so that where appropriate numbers are written as powers of ten. For example, 1,000,000 may be written as 10^6, and 3,500,000 as 3.5×10^6. Numbers smaller than one are indicated by negative powers: thus 0.00035 is written as 3.5×10^{-4}. In addition a variety of prefixes is used to denote certain multiples of units (*see* Table 1). Table 2 gives the SI equivalents of common imperial units while Table 3 lists a selection of astronomical constants.

Table 1: SI prefixes

Factor	Name	Prefix Symbol
10^{18}	exa	E
10^{15}	peta	P
10^{12}	tera	T
10^{9}	giga	G
10^{6}	mega	M
10^{3}	kilo	k
10^{2}	hecto	h
10^{1}	deca	da
10^{-1}	deci	d
10^{-2}	centi	c
10^{-3}	milli	m
10^{-6}	micro	μ
10^{-9}	nano	n
10^{-12}	pico	p
10^{-15}	femto	f
10^{-18}	atto	a

Table 2: SI conversion factors

Length	
1 in	25.4 mm
1 mile	1.609344 km
Volume	
1 imperial gal	4.54609 cm^3
1 US gal	3.78533 liters
Velocity	
1 ft/s	0.3048 $m\,s^{-1}$
1 mile/h	0.44704 $m\,s^{-1}$
Mass	
1 lb	0.45359237 kg
Force	
1 pdl	0.138255 N
Energy (work, heat)	
1 cal	4.1868 J
Power	
1 hp	745.700 W
Temperature	
°C =	kelvins − 273.15
°F =	$\frac{9}{5}$ (°C) + 32

Table 3: Astronomical and physical constants

Astronomical unit (A.U.)	1.49597870×10^8 km
Light-year (l.y.)	9.4607×10^{12} km = 63,240 A.U. = 0.306660 pc
Parsec (p.c.)	30.857×10^{12} km = 206,265 A.U. = 3.2616 l.y.
Length of the year	
Tropical (equinox to equinox)	$365^d.24219$
Sidereal (fixed star to fixed star)	365.25636
Anomalistic (apse to apse)	365.25964
Eclipse (Moon's node to Moon's node)	346.62003
Length of the month	
Tropical (equinox to equinox)	$27^d.32158$
Sidereal (fixed star to fixed star)	27.32166
Anomalistic (apse to apse)	27.55455
Draconic (node to node)	27.21222
Synodic (New Moon to New Moon)	29.53059
Length of day	
Mean solar day	$24^h03^m56^s.555 = 1^d.00273791$ mean solar time
Mean sidereal day	$23^h56^m04^s.091 = 0^d.99726957$ mean solar time
Earth's sidereal rotation	$23^h56^m04^s.099 = 0^d.99726966$ mean solar time
Speed of light in vacuo (c)	$2.99792458 \times 10^5\ km\,s^{-1}$
Constant of gravitation	$6.672 \times 10^{-11}\ kg^{-1}\,m^3\,s^{-2}$
Charge on the electron (e)	= 1.602 coulomb
Planck's constant (h)	$= 6.624 \times 10^{-34}$ J s
Solar radiation	
Solar constant	$1.39 \times 10^3\ J\,m^{-2}\,s^{-1}$
Radiation emitted	$390 \times 10^{26}\ J\,s^{-1}$
Visual absolute magnitude (M_v)	+4.79
Effective temperature	5,800 K

Contents

Foreword

Astronomy is the oldest of the sciences, born of the fact that man evolved on a planet from which he could see the sky. For 50,000 years he has had intelligence enough to study the heavens and attempt to understand what he saw by day and by night. And there is no doubt that what he saw and deduced from his observations has had extraordinary effects upon his life in many lands.

If our civilization had developed in the way it has on a planet whose skies were eternally cloud-covered (which is doubtful), we would have believed, up to some forty years ago, that the Earth *was* the universe. Only with the advent of large radar dishes, high-flying aircraft and rockets would the shattering fact have emerged that above the opaque cloud-layer lay a seemingly boundless universe.

Modern western civilization had been greatly influenced by Copernicus, Kepler, Galileo and Newton, all watchers of the skies. Our belief in a rational universe capable of being understood and our scientific and technological civilization spring from the cyclic behaviour of Sun, Moon, planets and stars and Newton's ability to explain so much of that behaviour by his law of gravitation and his three laws of motion. Timekeeping, navigation, geodesy, dynamics, religious and philosophical systems, cosmology and relativity and many other activities and interests of man have been directly affected by our study of the heavens.

There have been three astronomical revolutions. The first—the serious, naked-eye study of the heavens—lasted a long time—at least five millennia—and ended when Galileo began his systematic telescopic study of the sky in AD 1610. That second revolution, in which the camera and the spectroscope played their part, was brought to a climax by the enormous amount of information gathered by telescopes such as the 200-inch Hale telescope at Mt Palomar. On October 4, 1957, with the orbiting of Sputnik I, the third astronomical revolution began. Not only can we now place instruments in orbit above the Earth's atmosphere, obtaining access to the entire electromagnetic spectrum, but we can send spacecraft such as the Mariners, Pioneers and Voyagers to other planets in our Solar System. The flood of astronomical information has become a torrent, sweeping away many of our former ideas about the universe.

The time therefore seemed ripe for a series of atlases designed to take stock of this flood of new information and the new understanding it has brought us of the nature of the universe. Each atlas in the series has been written by an author chosen by Mitchell Beazley Publishers so that the text will provide the most up-to-date assessment of the celestial body studied, together with explanatory diagrams and the most modern pictures. Each author's text has then been carefully refereed by an acknowledged expert in the subject chosen by the Royal Astronomical Society's Education Committee. The final text of each book should therefore truly convey our present-day knowledge of the subject and remain a definitive work for many years to come.

Archie E. Roy
BSc, PhD, FRAS, FRSE, FBIS
Titular Professor of Astronomy in the
University of Glasgow
Chairman of the Education Committee of the
Royal Astronomical Society 1978–82

Introduction

The outermost planet known to the ancients was named by them Saturn, after the father of Jupiter. It was comparatively slow moving and shone with a dull, slightly yellowish light, which was regarded as baleful. Astrologically, Saturn's influence was regarded as unfavorable. There was no way in which observers of pre-telescopic times could tell that Saturn is, without doubt, the most beautiful object in the Solar System, indeed amongst all objects now visible to man in the sky.

Saturn's slow movement against the starry background means that it is well placed for observation in several months of each year. Any small telescope will show that there is something unusual about its form, and adequate magnification reveals the superb set of rings which distinguish Saturn from all the other planets in the Solar System. True, both Jupiter and Uranus have been found to have ring systems, but they can in no way compare with that of the planet Saturn.

Apart from Jupiter, Saturn is the largest and most massive planet in the Solar System. In constitution it is totally unlike the Earth; its outer surface is gaseous, and although several belts are always to be seen, the minor features change constantly. The surface is not as obviously active as that of Jupiter, but it is nevertheless in a state of constant turmoil, and recent space research methods have shown that the "winds" on Saturn are even more powerful than those of Jupiter. As with Jupiter, there are well-marked belts to either side of the equator, with less prominent belts in higher latitudes.

A large liquid planet

It would be wrong to suppose that Saturn is merely a smaller, ringed version of Jupiter. There are fundamental differences between the two giants, and these have been shown by the three probes that have so far bypassed Saturn: Pioneer 11, and Voyagers 1 and 2. Today it is believed that most of the globe is liquid, with only a comparatively small rocky core. In any event, Saturn is definitely a planet and not an embryo star. The core temperature is much too low to trigger off nuclear reactions.

Beautiful though it is, Saturn is not as consistently observed by amateur astronomers as is Jupiter. The main reason is that it shows fewer variations, apart from the changing tilt of the ring system. Yet violent outbreaks do occur: for instance, there was the famous white spot of 1933, discovered by an English amateur (W. T. Hay) and which provided important information about the planet's rotation period. Similar outbreaks may occur at any time, and at present (1982) no probes are monitoring Saturn, so that regular observations are extremely useful. It is, however, true that a telescope of adequate aperture is needed. Small instruments will show little apart from the main rings and belts. But an observer equipped with, say, a 30 cm reflector can play a valuable part in keeping the planet under regular surveillance.

The enigma of Titan

Quite apart from the planet itself, there is a particularly interesting system of satellites. One of these, Titan, lays claim to being the most fascinating world in the Solar System. It is the only planetary satellite to possess an atmosphere—and this atmosphere is quite different from what was expected. The Voyagers have shown that the main constituent is nitrogen, which makes up 78 percent of the Earth's atmosphere. Organic compounds exist, and there are all the ingredients for life, although the very low temperature has probably prevented life from evolving there. As yet we have no direct knowledge of the surface; the Voyagers could do no more than photograph the upper clouds, and we can hardly hope to learn more until a special probe is dispatched to orbit Titan and use radar to investigate the surface features. The other important satellites are smaller, though much larger than the junior members of Jupiter's family; they have icy surfaces, and are heavily cratered, although one of them, Enceladus, has a comparatively smooth surface and is the most reflective object yet to have been studied in the entire Solar System.

Before the Pioneer and Voyager missions, our knowledge of Saturn was relatively meager. We have learned a great deal, but there is still much that we do not know, and with the present cutbacks in NASA funds no further probes are planned. No doubt there will be new probes in the foreseeable future, both from America and from the Soviet Union. Meanwhile, Saturn remains as not only the most imposing but also one of the most interesting members of the Sun's planetary family.

1. The planet Saturn
Print from a color photograph by S. M. Larson, 11 March 1974.

2. Symbol of Saturn

3. The god Saturn

Saturn in the Solar System

Saturn, the most remote planet known in ancient times, is sixth in order of distance from the Sun. The ancients named the planet in honor of Jupiter's father, the first ruler of Olympus. Its mass is 95 times greater than that of the Earth—and is greater, in fact, than any other planet apart from Jupiter. It has more than 20 known satellites (*see* pages 54–86), of which one, Titan, is unique among satellites in the Solar System in having a dense atmosphere.

Saturn orbits the Sun at a mean distance of 9.538 astronomical units, equal to approximately 1,427 million km. Its orbit, like all the other planets, is appreciably eccentric: at perihelion (closest approach to the Sun) it lies at a distance of 1,347 million km, whereas at aphelion (furthest recession) it moves out as far as 1,507 million km, which is more than 10 times the distance between the Earth and the Sun. Saturn's orbital period is 10,759.2 Earth days (29.46 years). Like Jupiter, it has a rapid axial rotation: the currently accepted period is 10 hr 39 min 24 sec (appreciably longer than was believed before the flyby of Voyager 1). Saturn also exhibits differential rotation: that is, the period at the equator is shorter than that at higher latitudes. The quick spin means that Saturn is appreciably flattened at the poles. The equatorial diameter is 120,660 km, the polar diameter only 108,000 km.

Although Saturn is so massive (5.686×10^{26} kg), it is much the least dense of all the planets. Its mean density is only 0.69 gm/cm^3, which is less than that of the equivalent volume of water. Saturn's constitution, therefore, is quite unlike that of Earth. The outer layers are made up of gas, chiefly hydrogen and helium; the interior is mainly liquid hydrogen. There is a relatively small solid core, whose estimated diameter is 27,000 km. This is twice the diameter of the Earth, but because the core is so compressed, it is thought to have at least 15 times the mass of the Earth.

Brightness

As seen with the naked eye, Saturn is slightly yellowish in color and looks exactly like a stellar object, except that it twinkles less and can be seen, over long periods, to move against the stellar background. Star-gazers in ancient times, noting its rather dull light and sluggish motion, named it suitably; and, astrologically, Saturn was supposed to have an unfavorable influence. Yet, at its most brilliant, Saturn outshines most of the stars. Its maximum magnitude is −0.3, inferior only to two stars—Sirius and Canopus. This figure is bettered, however, by Venus, Jupiter, and Mars at its brightest.

The magnitude is affected by the angle of presentation of the ring system. In fact, the rings are more reflective than the planet: when they are edgewise-on to the Earth, as last happened in 1980, Saturn never becomes brighter at opposition than magnitude +0.8.

Oppositions

Saturn moves across the sky against the stellar background more slowly on average than any of the other naked-eye planets. The mean synodic period (the interval between successive oppositions) is 378 days, so that on average Saturn comes to opposition about a fortnight later each year: thus oppositions occurred on 1 March 1979, 13 March 1980 and 27 March 1981. The apparent diameter does not vary greatly: when it is at its closest to the Earth, Saturn's disc has an apparent diameter of 20.9 seconds of arc, which is reduced to 15 seconds of arc when the planet is furthest away.

The Rings

The glory of Saturn lies in its ring system (*see* pages 30–32). It is now known that both Jupiter and Uranus have rings, while Neptune may have, but those of Saturn, visible with a small telescope, are in a class of their own. There can be little doubt that Saturn is the most beautiful object in the entire sky.

There are three main rings. The two brightest (A and B) are separated by a gap known as the Cassini Division. This gap was until recently believed to be empty, but the Voyagers have shown a

1

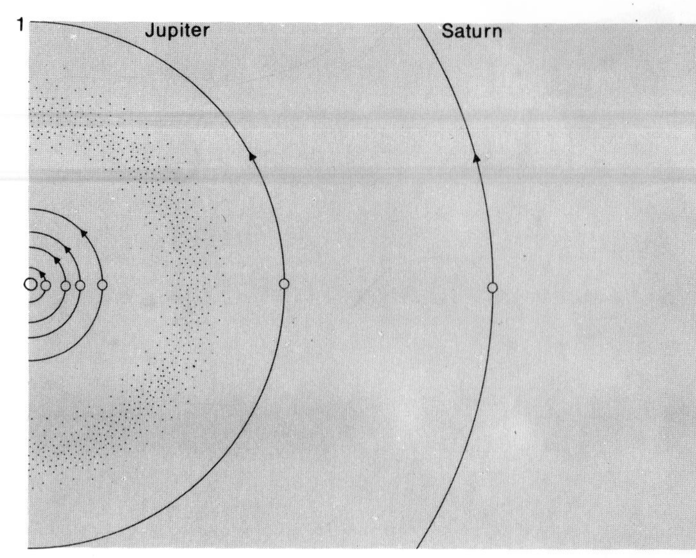

1. The Solar System
Nine major planets and many minor bodies are in orbit around the Sun. Saturn is the sixth planet from the Sun and the second giant planet beyond the asteroid belt (the speckled region in the diagram). The orbits of the inner planets—Mercury, Venus, Earth and Mars—are shown within the asteroid belt. Pluto's orbit is shown within that of Neptune, where it currently lies and where it will remain for approximately the next 20 years. Saturn's orbit, like those of all the planets, is an ellipse, with the Sun at one focus. Its mean distance from the Sun is 1,427 million km, more than nine times the distance of the Earth from the Sun.

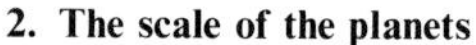

2

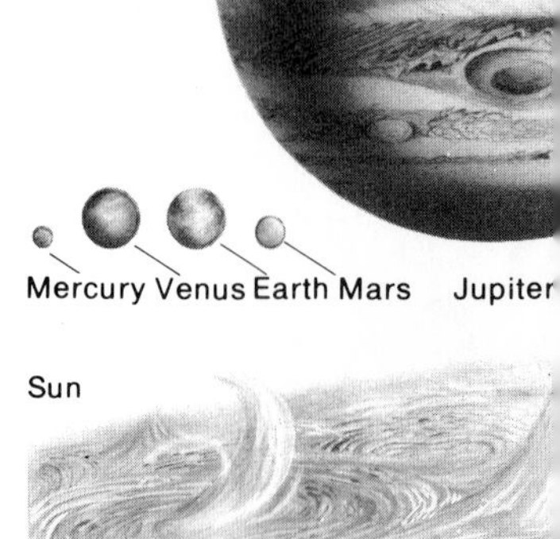

2. The scale of the planets
Saturn is the second largest of the planets, after Jupiter, and is drawn to scale in this diagram of members of the Solar System. Part of the limb of the Sun is included for size comparison.

3

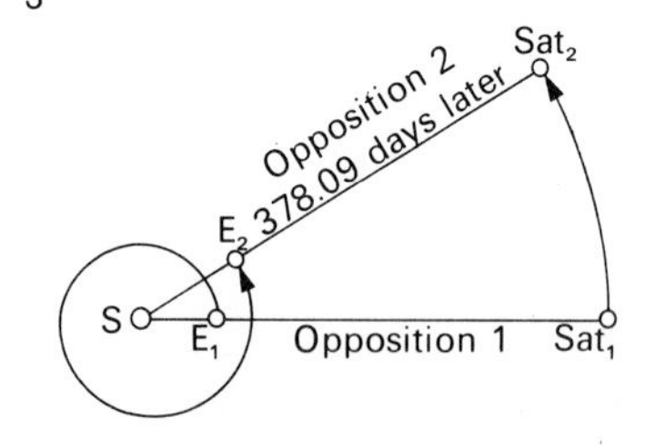

3. Oppositions
The configuration shown in the diagram, such that Saturn (Sat$_1$), the Earth (E$_1$) and the Sun (S) form a straight line, is called an "opposition". The next opposition will occur approximately 54 weeks later, when the Earth will have moved to position E$_2$ after one orbit around the Sun, and Saturn, which moves much more slowly, will have reached position Sat$_2$.

Mean orbital elements of Saturn and the Earth

Orbital elements	Earth	Saturn
Mean distance (astron units)	1.000000	9.53884
(millions of km)	149.6	1,427.0
Sidereal period (days)	365.256	10,759.22
Synodic period (days)	—	378.09
Eccentricity	0.0167	0.0556
Inclin. to the ecliptic (degrees)	0	2.49
Mean longitude of:		
Ascending node (degrees)	—	113
perihelion (degrees)	103	92

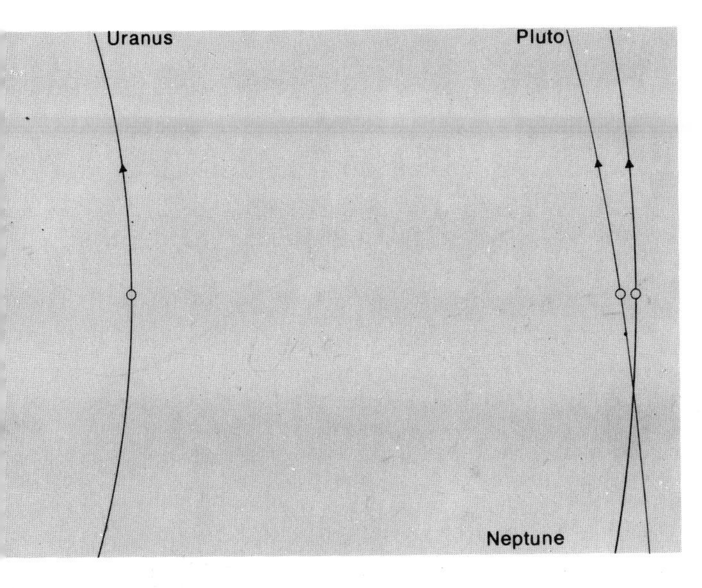

number of narrow rings within it. There is a gap in the outer ring (A) known as the Encke Division. The third famous ring, the C or Crêpe Ring, is closer to the planet than the other two, and is semitransparent.

Before the flights of Pioneer 11 and Voyager 1 these were the only rings definitely known to exist. A faint ring on the outside of the main system had been reported on various occasions, and also a ring (D) closer in than the Crêpe Ring, but the evidence was not conclusive. In fact, there are three rings—G, F and E—outside the main system. There is ring material closer in than the Crêpe Ring, although it does not seem likely that this could ever have been seen through an Earth-based telescope.

One of the greatest surprises of the Voyager mission was the discovery that the rings are much more complicated in structure than had been expected: there are hundreds of minor divisions. The divisions are not empty, and one of the outer rings has a curiously twisted or braided appearance, the reason for which is still not fully understood. There are also strange spoke-like features in the B Ring which were entirely unsuspected—indeed, could not have been imagined—before Voyager's flyby.

Spectra

Spectroscopic studies carried out from Earth were adequate enough to confirm that the composition of the atmosphere (*see* pages 22–25) is quite different from that of the Earth. Saturn, like all the planets, shines by reflected sunlight, so it yields what is basically a solar spectrum. The planet's own materials (*see* pages 18–19), however, leave their own imprint. Atoms or molecules emit or absorb radiation (depending on their temperature) and every substance emits or absorbs at a certain characteristic wavelength. This shows up as lines on the electromagnetic spectrum. It was no surprise to find that the chief constituent is hydrogen—together with hydrogen compounds such as methane and ammonia, and also a considerable amount of helium, which is hard to detect directly by means of the spectroscope. Dark absorption lines in the spectrum of Saturn were first detected by Angelo Secchi as long ago as 1863: bands due to methane and ammonia were identified by T. Dunham at Mount Wilson in 1932. There is more observable methane but less ammonia than on Jupiter, due to Saturn's lower temperature as a result of its greater distance from the Sun.

The Satellites

Saturn has a wealth of satellites. One of these, Titan, was discovered by C. Huygens in 1665, and until recently was believed to be the largest satellite in the Solar System. The Voyager results showed that it is slightly smaller than Ganymede in Jupiter's system. Titan is unique, however, in having a dense atmosphere, made up chiefly of nitrogen: "clouds" hide the surface completely. Of the remaining satellites, Iapetus and Rhea are about 1,500 km in diameter; Dione and Tethys about 1,100 km, and the rest considerably smaller. Apart from Titan, it seems that these satellites are made up mainly of ice, and most of them have cratered surfaces. Several display puzzling surface features.

The Pioneer and Voyager missions led to the discovery of several new satellites, some of which are of interest in as much as they have unusual orbits. The outer rings are "controlled" by small satellites, and there are several "co-orbital" satellites—that is, two or more satellites moving in the same orbit. The first of these to be discovered was a small satellite sharing the orbit of Dione, keeping 60° ahead of it.

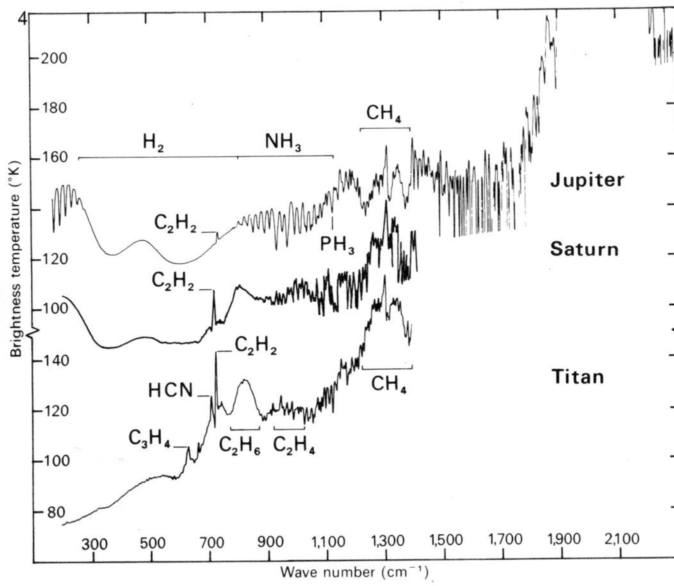

4. Spectra
Absorption and emission lines on a planet's spectrum identify the constituents of its atmosphere. Data from the spectra of Saturn, its satellite Titan, and Jupiter are compared on this graph, which plots the brightness temperature of the radiation received from each object against its wave number. In each case prominent spectral features, indicating the presence of certain important molecules, are revealed.

The Nomenclature

Through the telescope, Saturn's disc appears yellowish and obviously flattened. As with Jupiter, the dark streaks crossing the planet are known as belts and the bright bands as zones. Since the surface is gaseous, there can be no stable features, but several belts are always to be seen, and change little in latitude. The more stable features have been given a standard nomenclature (opposite).

In the past it has been normal practice to reproduce photographs and drawings of Saturn with south at the top and north at the bottom, which is how the planet appears in an ordinary astronomical telescope (to an observer in the northern hemisphere of the Earth); most astronomical telescopes give an inverted image. Latterly, however, the official procedure has been to put north at the top, and this is the practice followed in the book.

Similar confusion has been caused with regard to "east" and "west". With Saturn, this problem may be avoided by referring to the "preceding limb" (where features are carried out of view by virtue of the planet's rotation) and the "following limb" (where they first appear from the averted hemisphere). Thus with north at the top of the picture the preceding limb is to the right and the following limb to the left.

Principal features

The belts of Saturn are less prominent than those of Jupiter, partly because Saturn is smaller and further away, and partly because they are genuinely less pronounced. Saturn has a blander appearance because of the greater amount of haze overlying the cloud tops. However, the two main belts, the North Equatorial and the South Equatorial, are always present, and this applies also to the less pronounced North and South Temperate Belts. Observationally, however, these belts are not always on view, because when the rings are wide open they hide a considerable part of one hemisphere.

Rotation

As with Jupiter, various definite features have rotation periods of their own, so that they drift about in longitude; but well-marked spots observable from Earth are rare, so that information is much less complete than for Jupiter. For example, there is nothing on Saturn to be compared with Jupiter's Great Red Spot, and apart from the belts, there are no features that are permanent enough to be included in the official nomenclature.

There is a strong equatorial current on Saturn, as with Jupiter, and indeed the windspeeds appear to be considerably greater, but the two planets differ in one important respect; the boundaries of different currents closely follow the limits of the belts and zones with Jupiter, but not with Saturn. Rotation periods of occasional brilliant white spots have been determined, but the spots themselves do not persist for long enough to yield really satisfactory results.

The rings

The main rings of Saturn are, in order of decreasing distance, lettered A, B and C. The gap between A and B is the Cassini Division. The narrow division in the A Ring is known as the Encke Division. The many narrower divisions in the main system, dividing it into thousands of ringlets, are not observable from Earth, and to attempt to give them a separate nomenclature would be impossible. The Voyager results also show that narrow ringlets exist in both the Cassini and Encke divisions. Ring material between the Crêpe Ring (C) and the planet is now known officially as Ring D.

The Pioneer 11 and the Voyager missions have increased the knowledge of the ring system beyond all recognition, and the nomenclature has been officially established. There is a division between rings C and D, known as the French Division. Outside Ring B is Ring F, separated from Ring B by the Pioneer Division. Outside this is the very elusive G Ring, which is extremely hard to detect, and has been seen from Voyager only in forward-scattering light. Beyond this comes the E Ring, which is 90,000 km broad.

North Polar Region (NPR) Lat. +90° to +55° approx.
The northernmost part of the disc. Its color is variable: sometimes it is comparatively bright; at other times it is comparatively dusky in appearance.
North Temperate Zone (NTZ) Lat. approx. +70° to +40°
Generally fairly bright; but from Earth few details can be seen.
North Temperate Belt (NTB) Lat. +40°
One of the more active belts on the disc, and usually easy to see telescopically except when covered by the rings.
North Tropical Zone (NTrZ) Lat. +40° to +20°
A generally fairly bright zone between the two dark belts.
North Equatorial Belt (NEB) Lat. +20°
A prominent belt, always easy to see and generally fairly dark. Activity within it can sometimes be observed from Earth.
Equatorial Zone (EZ) Lat. +20° to −20°
The brightest part of the planet. Details can be observed in it, and there are occasional white spots. The most prominent example of white spots in the twentieth century was in 1933.
South Equatorial Belt (SEB) Lat. −20°
A dark belt, usually about the same intensity as the corresponding belt in the northern hemisphere.
South Tropical Zone (STrZ) Lat. −20° to −40°
A generally bright zone. Little detail to be seen telescopically.
South Temperate Belt (STB) Lat. −40°
Generally visible when not covered by the rings, but it is seldom really prominent.
South Temperate Zone (STZ) Lat. −40° to −70°
A brightish zone, with little or no visible detail as seen from Earth.
South Polar Region (SPR) Lat. approx. −70° to −90°
The southernmost part of the disc. Like the north polar region, somewhat variable in its depth of shading.

Saturn is predominantly liquid so it does not spin in the same way as a solid body, and features at different latitudes on the planet rotate in different periods. Conventionally, as with Jupiter, there are two rotation periods for Saturn, known as System I and System II. The region between the two equatorial belts is termed System I and this has a rotation period that is appreciably more rapid than that of the rest of the planet (System II). The two rates provide a fairly accurate guide to the relative movement of different features against which longitude can be measured.

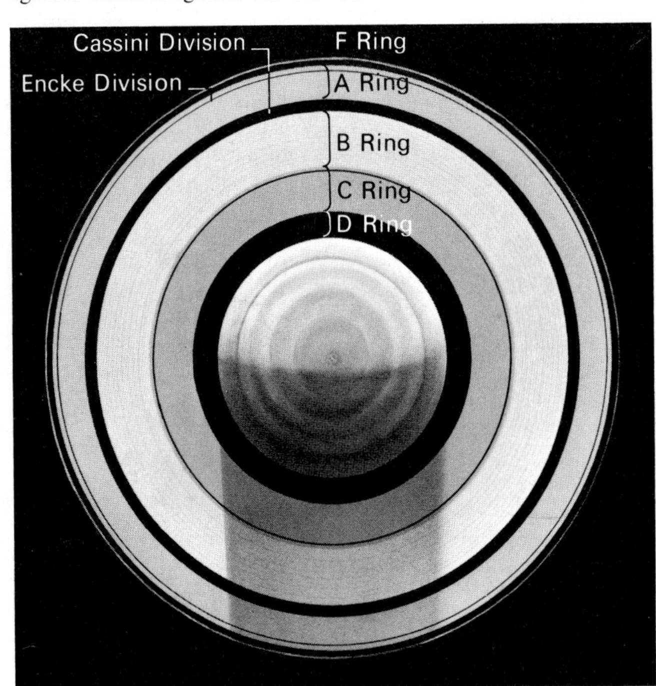

Early Observations

Saturn must have been known since very early times because, at its brightest, it outshines all stars except Sirius and Canopus. The first recorded observations of Saturn seem to have been made in Mesopotamia in the mid-7th century BC. About 650 BC there is a record that Saturn "entered the Moon", which is presumably a reference to an occultation of the planet. But it was not until July 1610 that Galileo turned his telescope towards Saturn, and saw the planet's disc for the first time.

The rings: nature and structure

Galileo was puzzled by Saturn. The telescope he used had a magnification of only 32, but it was still powerful enough to show that there was something unusual about the planet's shape. In 1610 the rings were placed at a narrow angle to the Earth, and Galileo could not see them in their true guise; instead he wrote that "the planet Saturn is not one alone, but is composed of three, which almost touch one another and never move nor change with respect to one another. They are arranged in a line parallel to the Zodiac, and the middle one is about three times the size of the lateral ones."

Two years later a fresh surprise awaited him: the attendant globes had disappeared. What had happened, of course, was that the rings had become edgewise-on to the Earth, and Galileo had no hope of seeing them. He then wrote: "Are the two lesser stars consumed after the manner of solar spots? Has Saturn, perhaps, devoured his own children? . . . The unexpected nature of the event, the weakness of my understanding, and the fear of being mistaken, have greatly confounded me."

He never solved the mystery, though when he looked at Saturn again, in 1616, the rings were wider open, and had he been seeing them for the first time he might well have realized the truth. A drawing made by the French astronomer P. Gassendi, in November 1636, also shows what seems to be a ring, but Gassendi too failed to interpret it correctly, and the true explanation was not given until 1659, in Christiaan Huygens' celebrated *Systema Saturnium.* Huygens had begun his telescopic observations in 1655, with a telescope that would bear a magnification of 50, and in his book he gave the solution of an anagram which he had published earlier to ensure priority. The anagram read, in translation: "[The planet] is surrounded by a thin flat ring, nowhere touching [the body of the planet] and inclined to the ecliptic."

Earlier explanations of the rings seem strange today. For example, the French mathematician Gilles de Roberval believed Saturn to be surrounded by a torrid zone giving off vapors, transparent and in small quantity but reflecting sunlight at the edges if of medium density, and producing an elongated appearance if very thick. Another French mathematician, Honoré Fabri, believed that the appearances could be explained by the presence of two large, dark, unreflecting satellites close to the planet and two large bright ones further away. Another theory was proposed in 1658 by Sir Christopher Wren, who had been a professor of astronomy at Oxford before turning to architecture. Wren believed that Saturn had an elliptical corona, meeting the globe in two places and rotating with Saturn once in each sidereal period. Wren never published his theory, because before he was ready to do so he heard of Huygens' solution, and accepted it at once. Others, including Fabri, were less wise. However, further observations showed that Huygens' theory was incontestable, and by 1665 even his most vehement opponents had accepted it.

Divisions in the rings

In 1675 the Italian astronomer G. D. Cassini, who had become the first Director of the Paris Observatory, found a dark line in Saturn's ring which subsequently proved to be in the nature of a gap, and is named in his honor. Up till then it had been tacitly assumed that the rings must be either solid or liquid, but the discovery of a division cast obvious doubt upon this theory, though the first definite

1

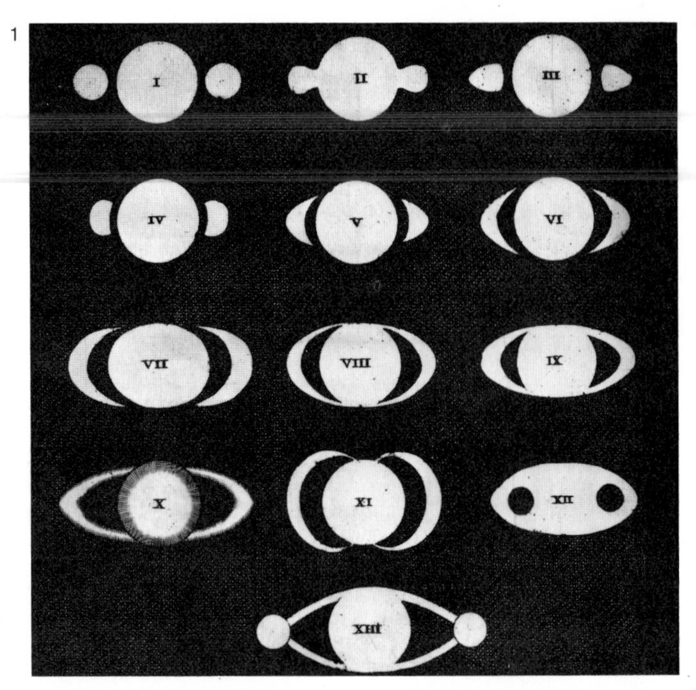

1. Early drawings of Saturn
Taken from Huygens' *Systema Saturnium* they are: I Galileo (1610); II Scheiner (1614); III Riccioli (1614 or 1643); IV–VII Hevel (theoretical forms); VIII and IX Riccioli (1648–50); X Divini (1646–8); XI Fontana (1636); XII Biancani (1616); XIII Fontana (1644–5). Some of these drawings, notably IX, had a very ring-like appearance several years before Huygens' theory was widely accepted.

2. Huygens' ring cycle
In *Systema Saturnium*, Huygens explained the inclination of Saturn's rings according to the planet's orbital position with respect to the Earth. As a result of Saturn's axial inclination, the rings are at maximum opening only at A and C, and then, Huygens stated, Saturn is at its most brilliant. The tilt of the planet also results in a pronounced seasonal variation, as is experienced on Earth.

2

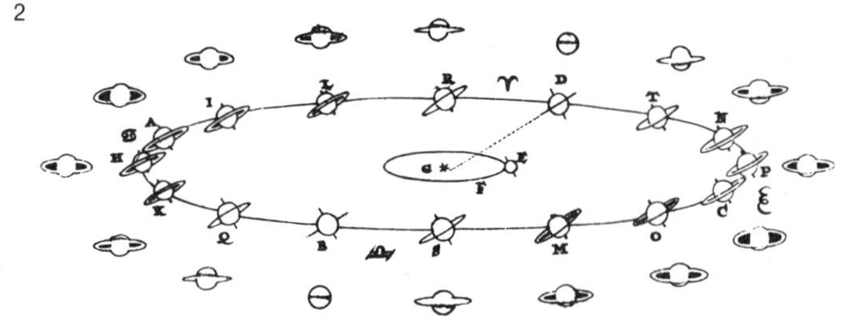

3. The Cassini Division
This sketch of Saturn, made in 1676 by J. D. Cassini, shows the division in the ring. This was the first known drawing in which the Division is unmistakably shown, and proves that credit for its discovery must go to Cassini.

3

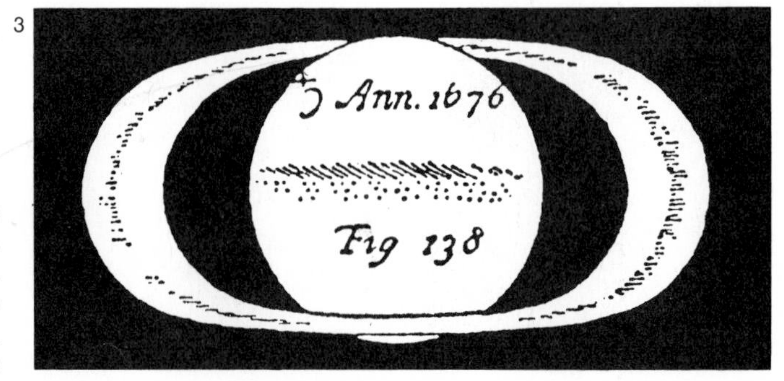

challenge to it was delayed until 1705, and was due to J. J. Cassini, who had succeeded his father in Paris.

A division in Ring A was announced by J. F. Encke, Director of the Berlin Observatory, in 1837. It was much less prominent than the Cassini Division, and for some time its nature was regarded as dubious, though there could be no doubt that a feature of some sort existed there. Minor divisions in both bright rings were also reported by various observers, but confirmation was lacking, and it is now known that the many minor divisions which actually exist cannot possibly be seen from Earth.

The Crêpe Ring

The next major advance came in 1850, with the discovery of Ring C. It was due to W. C. and G. P. Bond, at Harvard, although it seems that the first true interpretation of their observations was made by C. W. Tuttle, an assistant at the Harvard Observatory. Quite independently the ring was discovered by W. R. Dawes in England, and was at once confirmed by W. Lassell. In 1852 Lassell found that the ring was more or less transparent. It is, in fact, not a difficult object and it was suggested that the ring was not observed earlier because it was fainter. This, however, is improbable.

The Bonds regarded the rings as fluid, and this was also the opinion of B. Peirce, who published his theories in 1855. But this posed problems, and in 1855 the University of Cambridge announced that the subject of their Adams Prize Essay would be "to determine the extent to which the stability and appearance of Saturn's rings would be consistent with alternative opinions about their nature—whether they are rigid or fluid, or in part aeriform". The Prize was won in 1857 by James Clerk Maxwell, who showed that neither a solid, a liquid nor a gaseous ring could persist. He concluded, therefore, that the rings were nothing more nor less than swarms of small particles, so close together that at the distance of Saturn they gave the impression of a solid sheet.

Final proof came in 1895 by J. E. Keeler, who showed spectroscopically that the inner sections of the ring system revolve around the planet faster than the outer parts, in agreement with Kepler's Laws. By means of the Doppler effect, Keeler was even able to determine the velocities of the ring particles. His investigation was quickly followed up by W. W. Campbell, who gave the following velocities: for the inner edge of Ring B, 18.94 km/sec; middle of the bright ring, 17.37 km/sec; and outer edge of Ring A, 15.8 km/sec.

The Kirkwood Gaps

The next important investigation was due to the American astronomer Daniel Kirkwood, in 1866, when he found that there are well-marked gaps in the zone of minor planets or asteroids, which move around the Sun between the orbits of Mars and Jupiter. There are certain "zones of avoidance" in which the revolution periods would be simple fractions of the period of Jupiter (11.75 years); cumulative perturbations would, therefore, drive any asteroids out of this forbidden region. In the following year Kirkwood applied the same theory to Saturn's rings, and concluded that the Cassini Division was due to the cumulative perturbations upon ring particles produced by the inner satellites Mimas, Enceladus, Tethys and Dione. In the Cassini Division, a ring particle would have a period half that of Mimas, one third that of Enceladus, one quarter that of Tethys and one sixth that of Dione, so that it would soon be moved out of the Division.

The space-probe results have since shown, however, that although the effect may play an important role in the production of the Cassini Division, it cannot possibly be the complete answer in view of the many minor divisions now known and, even more significantly, the fact that there are several well-defined ringlets inside the Cassini and Encke divisions.

An occultation of a star (or Saturnian satellite) by the ring system gives valuable information concerning the densities of the various rings. Important observations were made in 1917 by M. A. Ainslie and J. Knight when the star BD +21°1714 was occulted. These confirmed that Ring C is transparent and Ring A partly so. This principle was used much more effectively in August 1981, when Voyager 2 observed the occultation of the star Delta Scorpii by the ring system. Not only the main rings and divisions, but also the minor divisions and the ringlets were studied in this manner.

4

4. Johann Franz Encke
In 1837, Johann Encke of the Berlin Observatory saw Ring A divided by a black line. He obtained an accurate measurement of its position, which he described as on the inner part of the ring at approximately $\frac{1}{3}$ of the distance from the inner to the outer edge. The Encke Division was thus first seen.

5

5. Saturn's third ring
In 1850 at Harvard, G. P. Bond drew a new ring crossing the globe. The C Ring, later named the Crêpe Ring by William Lassell, was on the inside of the inner edge of the B Ring, and was dusky in appearance.

6

6. James Clerk Maxwell
James Clerk Maxwell was a Scottish theoretical physicist who showed that the rings of Saturn must be composed of millions of small satellites.

7

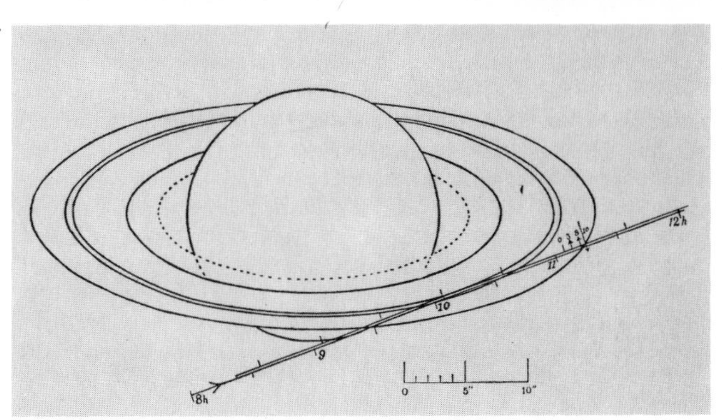

7. Occultation of BD +21°1714
A faint star, BD +21°1714, passed behind the rings on 9 February 1917. The occultation was observed by M. A. Ainslie, and the drawing shows the path of the star, which grazed the disc and was obscured by the rings.

Early Theories

In general, early observers tended to concentrate upon the ring system of Saturn rather than on the disc of the planet. This is not surprising, particularly as the disc is so lacking in detail when compared with Jupiter. Belts were observed: the first definite observation of a belt was probably made by the Italian astronomer Campani as long ago as 1664. Other features, however, were extremely elusive. The first definite observations of spots were made at Johann Schröter's observatory at Lilienthal, near Bremen, by Schröter and his assistant Karl Harding in 1796, but rotation periods were very hard to determine with any accuracy.

Shape of the globe

The flattened appearance of Saturn is obvious with even a small telescope, and it is rather surprising that the first definite reference to it was not made until 1789, when William Herschel gave the ratio of the equatorial to the polar diameter as 11:10, which is approximately correct. In 1805 Herschel made a series of micrometrical measurements, and concluded that in addition to the polar flattening there was also an equatorial flattening, giving the planet a kind of square-shouldered appearance. It is now clear that he was deceived by a combination of the ring position, the polar shading and the low altitude of Saturn at the time of his observations.

Herschel also noted slight irregularities in the shape of the ring shadow across the globe. Schröter, in 1792, described these shadows as notched and jagged, and this was also the view of a leading English observer, William Lassell, in 1849. As it was generally believed at that time that the rings were solid, it was natural to assume that the shadow irregularities were due to real irregularities in the ring surface. In fact, the observations were due to atmospheric tremors and instrumental defects. It would be premature to claim that the rings are completely uniform, but any irregularities are certainly too slight to be seen from Earth.

Features of the disc

The disc of Saturn is not an easy subject for telescopic study, and well-marked features are uncommon. The most prominent spot of recent times was discovered on 3 August 1933 by the English amateur astronomer W. T. Hay, using the 15.2 cm refractor at his private observatory in London. The spot was a conspicuous object, and on 9 August Hay found that it took 51 minutes to cross the central meridian. It gradually lengthened and became less distinct; by 13 September it could no longer be called a true spot, but merely a somewhat brighter region of the equatorial zone. Other white spots have been seen since, but all are much more short-lived than the features on Jupiter.

Large telescopes were not often turned towards Saturn before the era of the space-probe and those that were made very little contribution to our general knowledge of the planet. In 1954 G. P. Kuiper made some observations with the Palomar reflector. He showed little disc detail, and considered that the Cassini Division was a genuine division, the Encke Division being a "ripple".

The nature of Saturn

Since it was known, from the early days of telescopic observation, that Saturn and the other outer planets were quite unlike the Earth, being much larger and with gaseous surfaces, it was natural to assume that they might emit an appreciable amount of heat, so that they would almost qualify as miniature stars. This was summed up in 1882 by the English astronomer R.A. Proctor, who published the first book of near-modern times to be devoted entirely to the planet: *Saturn and its System*. He wrote:

"Regarding the cloudphenomena of the giant planets as generated by internal forces, whose real seat lies deep below the visible surface of the cloud belts, we see that these forces must be of tremendous energy, must produce enormous changes in the cloud-laden atmosphere (with effects extending widely, both vertically and laterally), and imply enormous heat in the whole frame of each planet. . . . Over a region hundreds of thousands of square miles in extent, the flowing surface of the planet must be torn by sub-planetary forces. Vast masses of intensely hot vapor must be poured forth from beneath, and, rising to enormous heights, must either sweep away the enwrapping mantle of cloud which had concealed the surface, or must itself form into a mass of cloud."

This theory was still taken seriously until the work of Sir Harold Jeffreys, who published several important papers in the 1920s, the first in 1923. Jeffreys pointed out that the temperatures of the surface, which had been determined by D. H. Menzel, were too low to be reconciled with the miniature-sun theory. In a further paper, published in 1926, he demonstrated that a much more likely theory involved a rocky core, covered by a thick layer of ice, which was surrounded by an atmosphere with a depth amounting to about 23 percent of the radius of the planet.

Since hydrogen is the most abundant element in the universe, it was natural to assume that Saturn must contain a great quantity of it, and this was supported in 1932, when T. Dunham at the Mount Wilson Observatory, used the spectroscope to detect methane and some ammonia in the Saturnian atmosphere; both these are hydrogen compounds. This led R. Wildt, in 1938, to propose a modification of Jeffreys' theory, though retaining the essential characteristics of it: a rocky core, a layer of ice, and then a hydrogen-rich atmosphere. A rather different model was proposed in 1951 by W. Ramsey, who considered that hydrogen was the main constituent of Saturn, but that near the center the immense pressures forced it to assume many of the characteristics of a metal.

The current theory—that Saturn has indeed a silicate core, but that most of the hydrogen is liquid—was proposed before the space missions were dispatched, and now seems to be well established. Though the core of Saturn is undoubtedly hot and Saturn does emit radiation, the outer layers are extremely cold. Our views of Saturn have changed beyond all recognition since Proctor wrote his book a century ago.

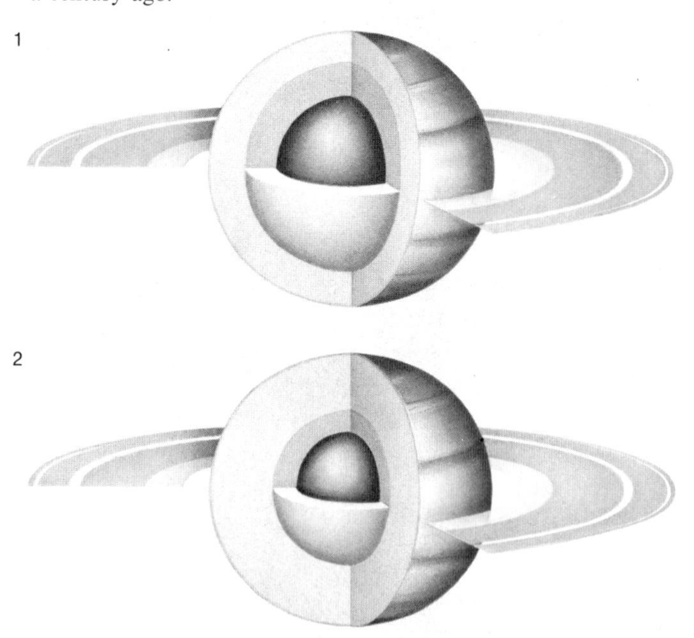

1. Jeffreys' model of Saturn, 1923
Jeffreys dismissed the belief that the giant planets were very hot and gaseous, and said they were cold and solid, but composed of low-density materials—hydrogen, nitrogen, oxygen, helium and perhaps methane.

2. Wildt's model of Saturn, 1938
Wildt assumed a metallic and rocky core, with a density of 6 gm/cm^3. The surrounding layer of frozen water and carbon dioxide (1.5 gm/cm^3) was topped by solid hydrogen (0.3 gm/cm^3) and finally the atmosphere.

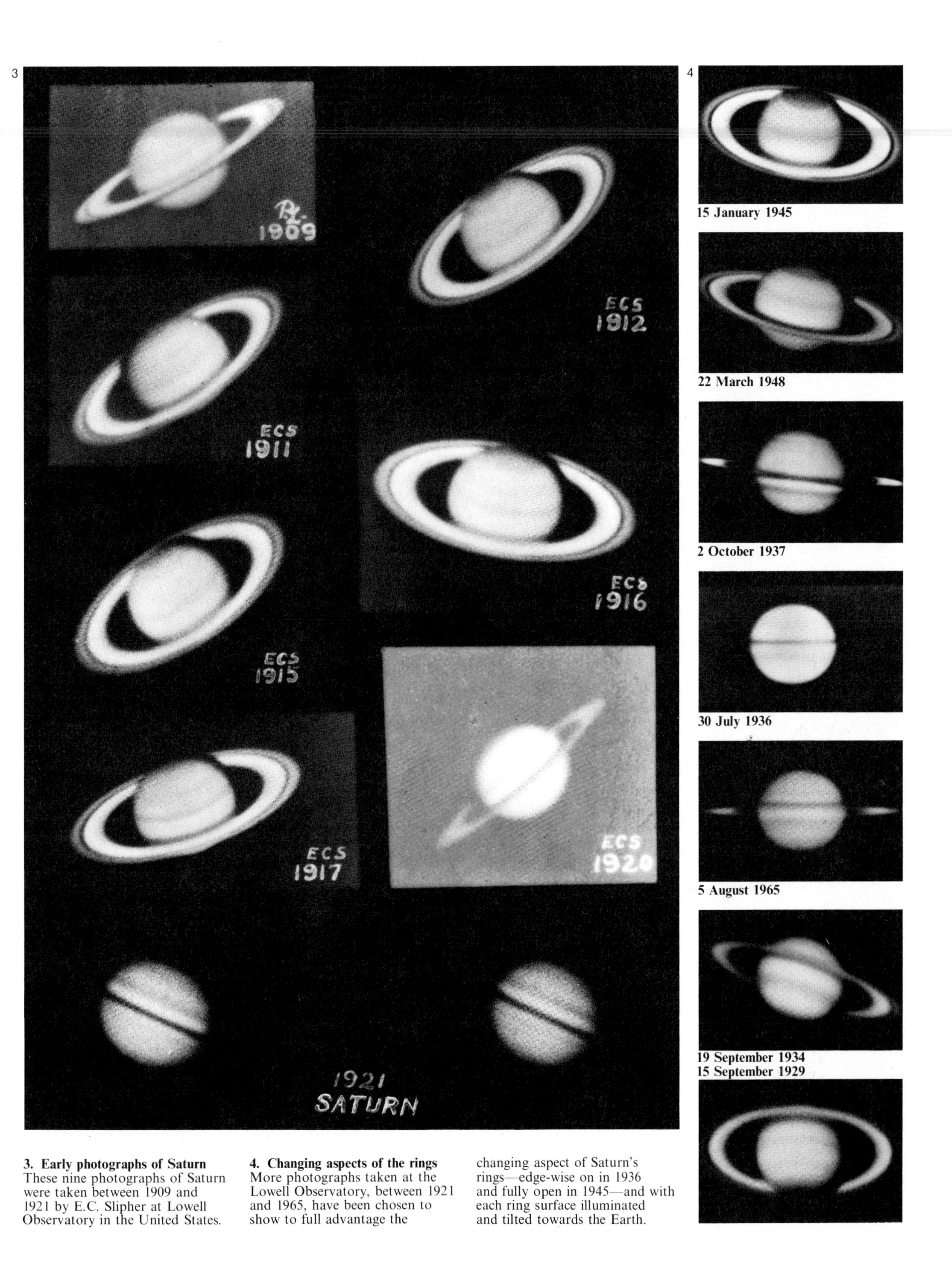

15 January 1945

22 March 1948

2 October 1937

30 July 1936

5 August 1965

19 September 1934

15 September 1929

3. Early photographs of Saturn
These nine photographs of Saturn were taken between 1909 and 1921 by E.C. Slipher at Lowell Observatory in the United States.

4. Changing aspects of the rings
More photographs taken at the Lowell Observatory, between 1921 and 1965, have been chosen to show to full advantage the changing aspect of Saturn's rings—edge-wise on in 1936 and fully open in 1945—and with each ring surface illuminated and tilted towards the Earth.

The Voyager Spacecraft

Three vehicles have so far bypassed Saturn. The two most recent, Voyager 1 and Voyager 2, are more advanced in design than their predecessor, Pioneer 11, in a considerable number of respects. In particular, their onboard computer systems are capable of directing more sophisticated experimental equipment, and the vehicles carry a more powerful source of electricity in the form of three Radioisotope Thermoelectric Generators (RTGs). While in flight, the spacecraft is stabilized along three axes, using the Sun and the star Canopus as celestial references. After the first 80 days of flight the 3.66 m parabolic reflector points constantly back to Earth.

Behind the reflector dish is a ten-sided aluminium framework containing the spacecraft's electronics. This framework surrounds a spherical tank containing hydrazine fuel used for maneuvering.

Twelve thrusters control the spacecraft's attitude, while another four are used to make corrections to the trajectory. Only trajectory changes require instructions from the ground; all other functions can be carried out by the onboard computer, in contrast to the Pioneer spacecraft which had be be flown "from the ground". There are three engineering subsystems: the Computer Command Subsystem (CCS), the Flight Data Subsystem (FDS) and the Attitude and Articulation Control Subsystem (AACS).

The entire vehicle weighs only 815 kg and carries equipment for 11 science experiments.

1

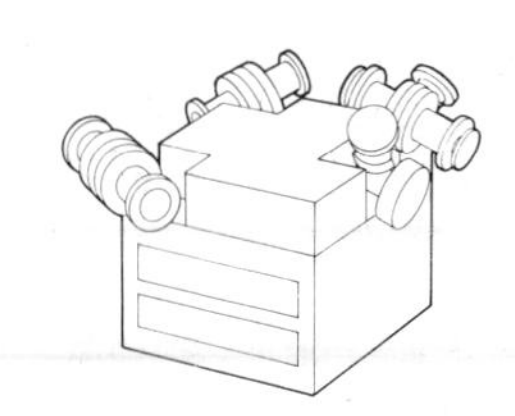

1. CRS
The Cosmic Ray Detector System is designed to measure the energy spectrum of electrons and cosmic ray nuclei. The instrument studied the composition of Saturn's radiation belts, as well as the characteristics of energetic particles in the outer Solar System generally. The experiment uses three independent systems: a High-Energy Telescope System (HETS), a Low-Energy Telescope System (LETS) and an Electron Telescope (TET). These enabled the spacecraft to study a wide range of particles that make up cosmic rays.

2

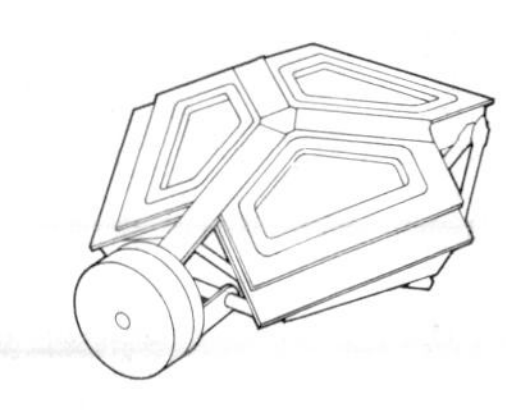

2. PLS
The Plasma experiment studied the properties of the very hot ionized gases that exist in the interplanetary regions. The instrument consists of two plasma detectors, one pointing in the direction of the Earth and the other at a right angle to the first. This equipment analyzed the properties of the solar wind and its interaction with first Jupiter and then Saturn. The PLS also studied the properties of Jupiter's and Saturn's magnetospheres, It is hoped that these studies will be extended to the planets still to be visited by Voyager.

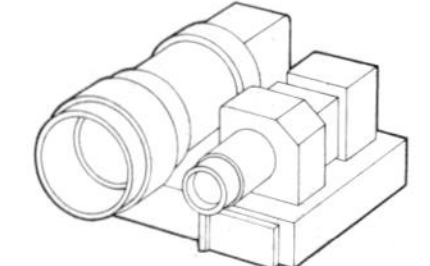

3. ISS
The Imaging Science Subsystem consists of two television-type cameras mounted on a scan platform. One of the cameras has a 200 mm wide-angle lens with an aperture of f/3, while the other uses a 1,500 mm f/8.5 lens to produce narrow-angle images. The design is a modified version of the cameras used on previous Mariner vehicles. Both cameras have a range of built-in filters as well as variable shutter speeds and scan rates. The ISS, the IRIS and the PPS instruments were able to view the same region of the planet simultaneously.

4. IRIS
The Infrared Radiometer Interferometer and Spectrometer measured radiation in two regions of the infrared spectrum, from 2.5 to 50 μm and from 0.3 to 2.0 μm. It provided information about the temperatures and pressures at various levels of Saturn's atmosphere, as well as about the chemical composition of its clouds. Mounted on a scan platform, the instrument has two fields of view, one using a 0.5m Cassegrain telescope to achieve a narrow, quarter-degree field of view, the other, pointed off the telescope sight, for a wider view.

5

5. LECP
The Low-Energy Charged Particle experiment uses two solid-state detector systems mounted on a rotating platform. The two subsystems consist of the Low Energy Particle Telescope (LEPT) and the Low Energy Magnetospheric Particle Analyzer (LEMPA). This equipment studied Saturn's magnetosphere (*see* page 20) and the interaction of charged particles with Saturn's satellites. It is also designed to investigate various other interplanetary phenomena such as the solar wind and cosmic rays emanating from sources outside the Solar System.

6

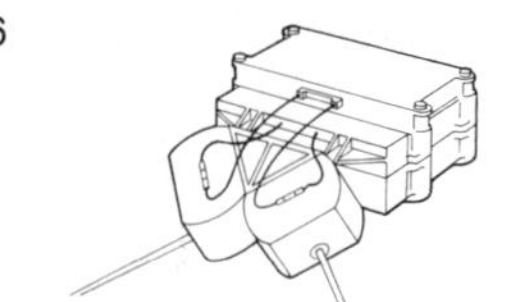

6. PWS and PRA
Two separate experiments, the Plasma Wave System and the Planetary Radio Astronomy experiment, share the use of the two long antennas which stretch out at right angles to one another forming a "V". The PWS studied wave-particle interactions and measured electric-field components of plasma waves over a frequency range of 10 Hz to 56 kHz. This system is also designed to measure the density of thermal plasma near the planets visited by the space-probe. The PRA experiment detected and analyzed radio signals emitted by the planets; Saturn was expected to be a powerful source of radio waves from Earth-based experiments. The PRA receiver covers two frequency bands, the first in the range of 20.4 kHz to 1,300 kHz, and the second between 2.3 MHz and 40.5 MHz.

7. PPS
The Photopolarimeter System consists of a 0.2 m telescope fitted with filters and polarization analyzers and is mounted on a scan platform. It covers eight wavelengths in the region between 235 μm and 750 μm. The PPS measured gases in planetary atmospheres, examined particles present in the atmospheres, and searched the sky background for interplanetary particles. The PPS instrument is also designed to examine the surface texture and composition of the satellites of Jupiter and Saturn, and the atmosphere of Titan.

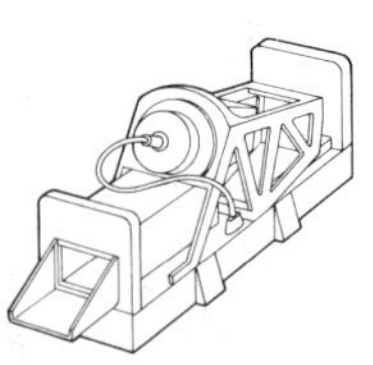

8. UVS
The Ultraviolet Spectrometer covers the wavelength range of 40 μm to 180 μm looking at planetary atmospheres and interplanetary space. Its purpose is to study the chemistry of the upper layers of the atmospheres, and to measure how much of the Sun's ultraviolet radiation they absorb during occultation; it also measures ultraviolet emissions from the planetary atmosphere. The instrument collects and channels light through a collimator, which directs a number of narrow parallel beams onto a diffraction grating.

9

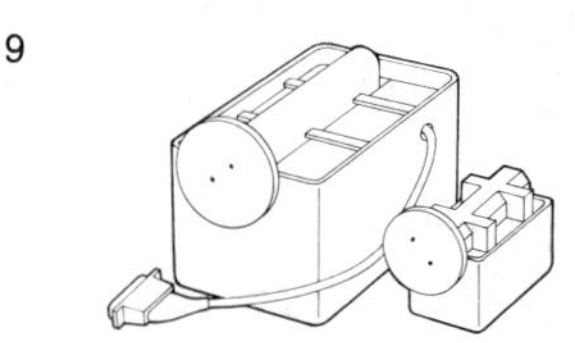

9. MAG
The Magnetic Fields Experiment consists of four magnetometers; two are low-field instruments mounted on a 10 m boom away from the field of the spacecraft, while the other two are high-field magnetometers mounted on the body of the spacecraft. Each pair consists of two identical instruments, which makes it possible to eliminate the spacecraft's field from the results. Each magnetometer measures the magnetic component along three perpendicular axes, from which the direction and strength of the field can be determined.

10. RSS
The investigations of the Radio Science System are based on the radio equipment which is also used for two-way communications between the Earth and Voyager. For example, the trajectory of the spacecraft can be measured accurately from the radio signals it transmits; analysis of the flight path as it passes near a planet or satellite makes it possible to determine the mass, density and shape of the object in question. The radio signals are also studied at occultations for information about the occulting body's atmosphere and ionosphere.

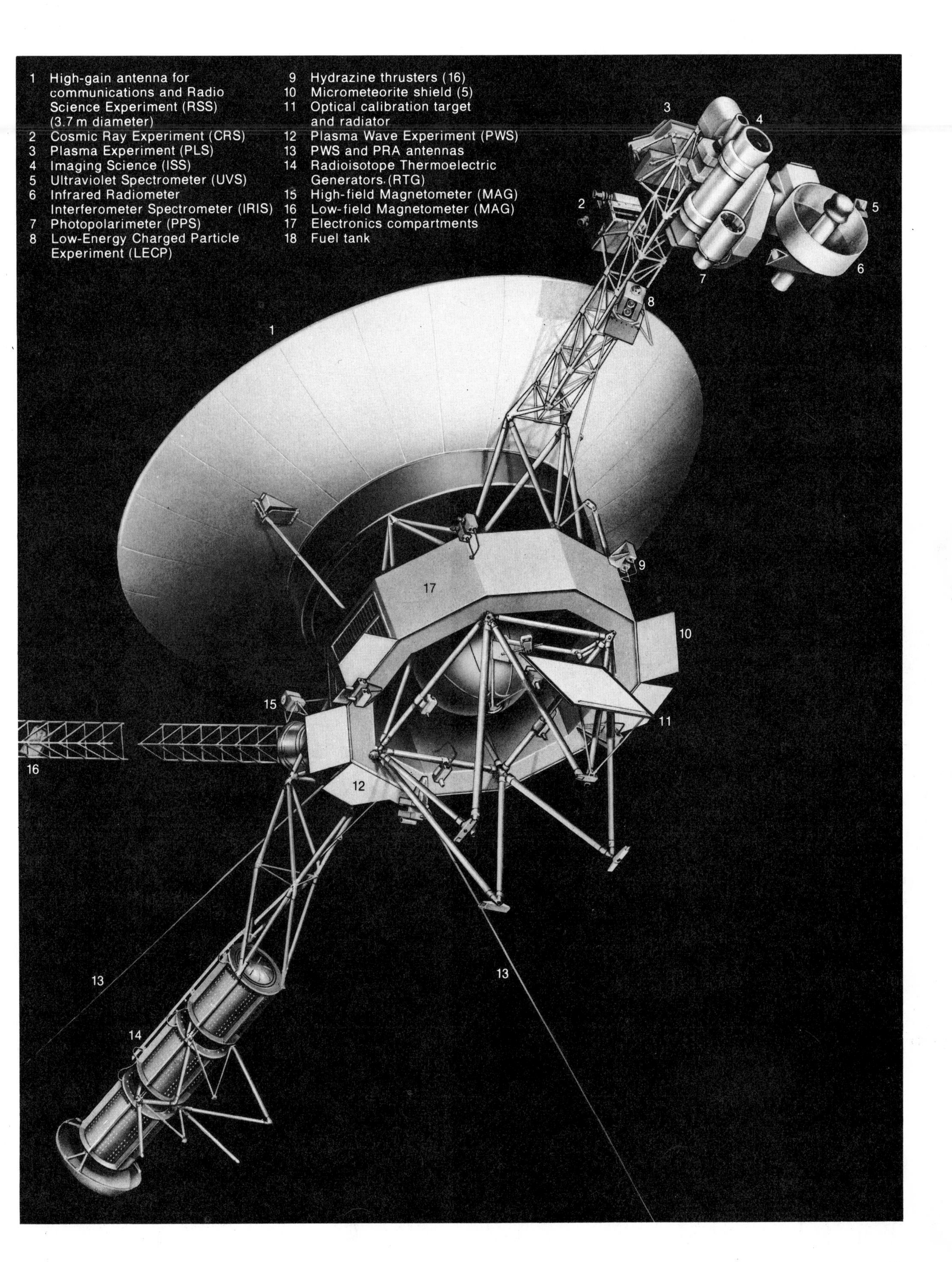
1 High-gain antenna for communications and Radio Science Experiment (RSS) (3.7 m diameter)
2 Cosmic Ray Experiment (CRS)
3 Plasma Experiment (PLS)
4 Imaging Science (ISS)
5 Ultraviolet Spectrometer (UVS)
6 Infrared Radiometer Interferometer Spectrometer (IRIS)
7 Photopolarimeter (PPS)
8 Low-Energy Charged Particle Experiment (LECP)
9 Hydrazine thrusters (16)
10 Micrometeorite shield (5)
11 Optical calibration target and radiator
12 Plasma Wave Experiment (PWS)
13 PWS and PRA antennas
14 Radioisotope Thermoelectric Generators (RTG)
15 High-field Magnetometer (MAG)
16 Low-field Magnetometer (MAG)
17 Electronics compartments
18 Fuel tank
1
2
3
4
5
6
7
8
9
10
11
12
13
13
14
15
16
17

Trajectories and Communications

The first spacecraft to reach Jupiter was Pioneer 10, launched on 2 March 1972, which made its closest approach to the planet on 3 December 1973 at a distance of approximately 132,000 km. Pioneer 11 was launched on 5 April 1973 and bypassed Jupiter on 2 December 1974, almost exactly a year after its predecessor.

Pioneers 10 and 11 were virtually identical, but Pioneer 11 made a closer approach to the planet (42,800 km) and passed more rapidly over the equatorial zone, following a trajectory which took it over Jupiter's south pole. It was then decided to make a course adjustment which would take Pioneer 11 back across the Solar System towards a rendezvous with Saturn on 1 September 1979. This extended program was successfully carried through, providing the first close-range information from Saturn. Both Pioneers are now travelling out from the Solar System, never to return.

In 1977 a rare alignment of the outer planets (once in 176 years) offered the possibility of sending space-probes on a "Grand Tour", taking them in turn past Jupiter, Saturn, Uranus and possibly Neptune as well. Because of the immense distances involved and the limited supply of fuel that can be carried on board, a special trajectory known as "gravity-assist" is used to boost the vehicle on its journey. The principle is to use the gravitational field of each planet to accelerate the spacecraft on to the next.

The two Voyagers were launched in the late summer of 1977. Voyager 2 was in fact launched first (20 August) and made its pass of Jupiter on 9 July 1979 at 714,000 km. It is scheduled to reach Saturn in August 1981, Uranus in January 1986 and possibly Neptune in 1989. Voyager 1, launched on 5 September 1977, travelled by a more economical route and reached Jupiter first, passing it at 350,000 km on 5 March 1979. Voyager 1 went on to reach Saturn in November 1980.

New probes have already passed the design stage. Of particular importance is Project Galileo, due to be launched in 1985, which is intended to send a probe into the Jovian atmosphere.

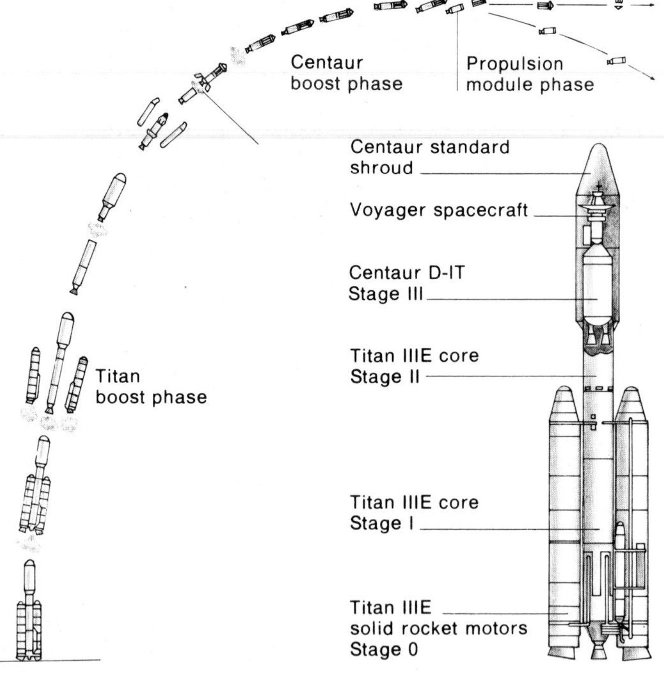

1. Titan/Centaur
The Voyager spacecrafts were launched from Cape Canaveral by a vehicle consisting of the Titan III-E and a Centaur D-1T upper stage, with a protective assembly called the Centaur Standard Shroud (CSS). The thrust at lift-off was provided by two Solid Rocket Motors (SRMs). The launch involved a total of six engine burns, the last provided by a propulsion module suspended below the mission module.

2. Voyager trajectories

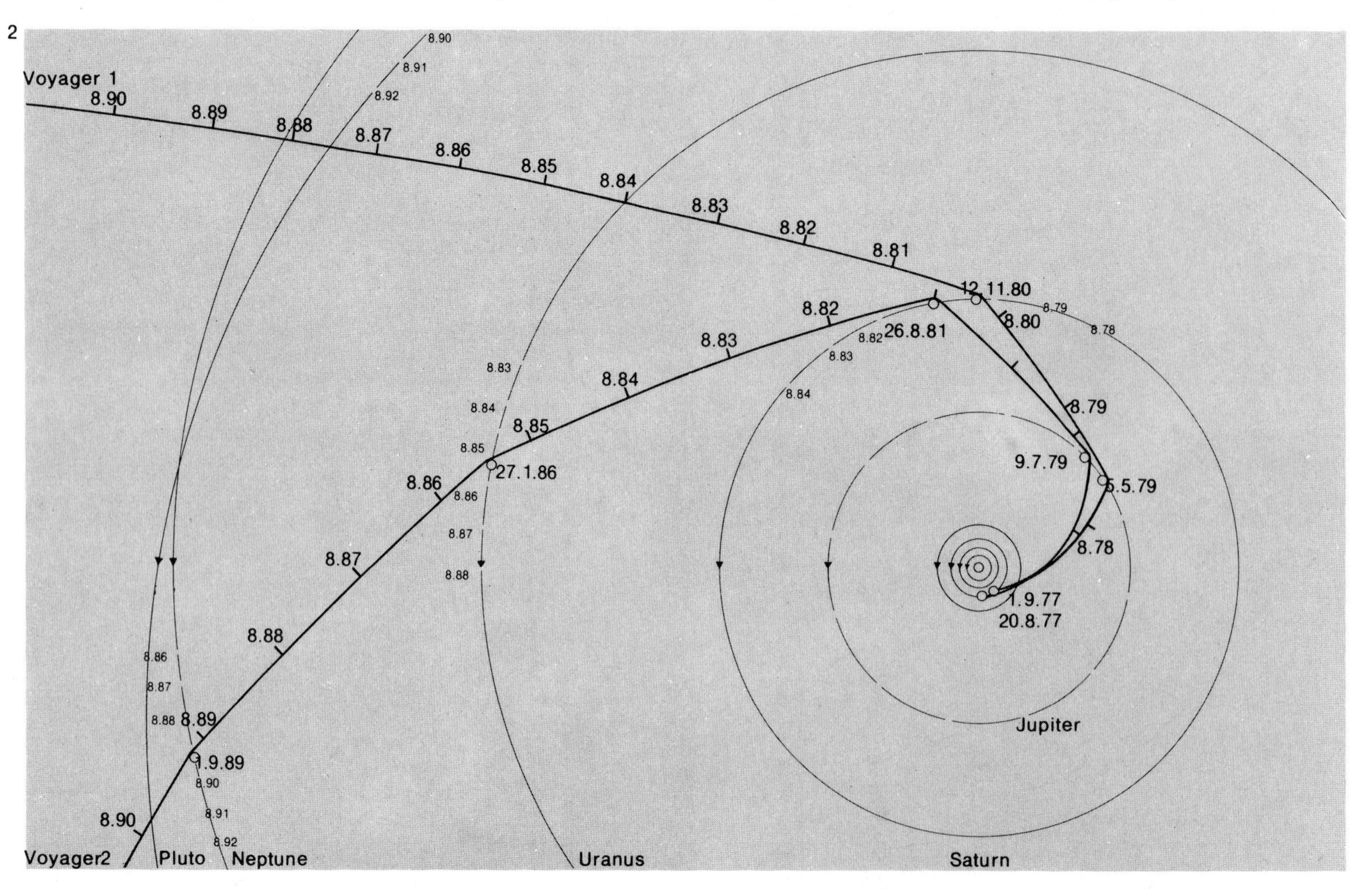

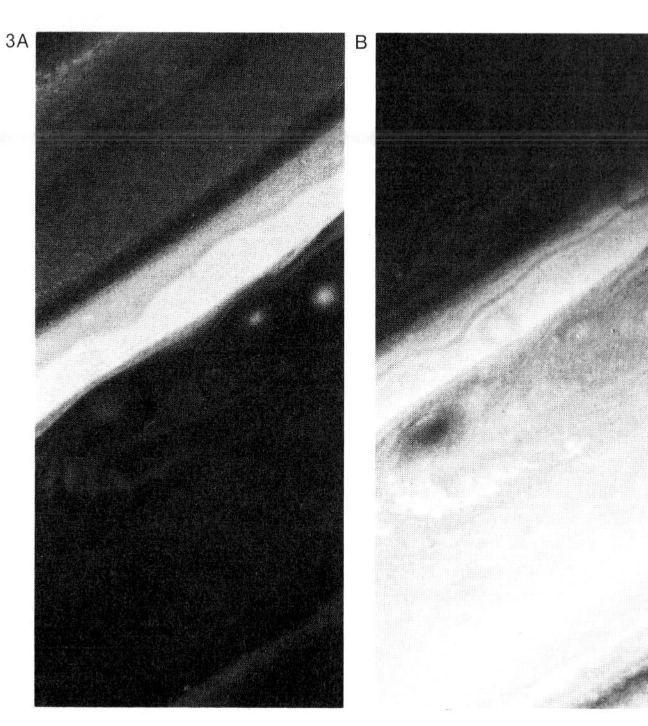

3. Image transmission
Full-color images are built up from monochromatic images taken through different filters. Color balance can be improved later as part of the computer processing to produce the final image. The violet image (**A**) shows three bright oval cloud systems, the largest of which appears dark in the green image (**B**) with a clockwise rotation.

4. Telecommunications system
This block diagram illustrates the stages of data transmission.

The problems involved in tracking, controlling and communicating with space-probes as far away as Jupiter are considerable. The power of the onboard transmitters is severely limited by the restriction on weight; those on the Voyager spacecraft operated on a maximum of less than 30 W, and the power of the received signal over 1 m^2 of the Earth's surface was in the range of only 10^{-18} W. Moreover, because of the immense distances, there is a delay of about 40 min between the transmission and reception of a signal.

Also, a certain amount of interference and distortion is inevitable. Some types of information, such as command signals, demand a high degree of accuracy; others, such as video data, are relatively tolerant, since errors can be removed by computer-processing on the ground. Sophisticated encoding techniques are used to protect the most error-sensitive data, while at the same time more robust data can be transmitted more economically. The communications system operated on two different frequencies, one near 0.13 m in the so-called "S-band" and another near 0.4 m—the "X-band". The X-band was used exclusively for "downlink" transmissions, while the S-band carried both "uplink" and "downlink" data. A network of three space stations, in Canberra, Madrid and Goldstone, tracked the Voyager spacecraft; these stations are sited at such longitudes that the spacecraft is within range of one at all times.

The transmission of images is of particular interest, and in this respect the Voyagers represent a significant advance over the Pioneers. In flight the Pioneer vehicles were constantly spinning about one axis, and their cameras built up an image gradually, scanning a narrow strip during each rotation. The Voyagers, on the other hand, employ shuttered television-type cameras. Each image is analyzed into 640,000 picture elements or "pixels"; the brightness of each pixel is measured, and its value expressed as an eight-digit binary number, thus converting the image into a long string of binary code that can either be transmitted immediately or stored in the memory, whose storage capacity is equivalent to 100 pictures.

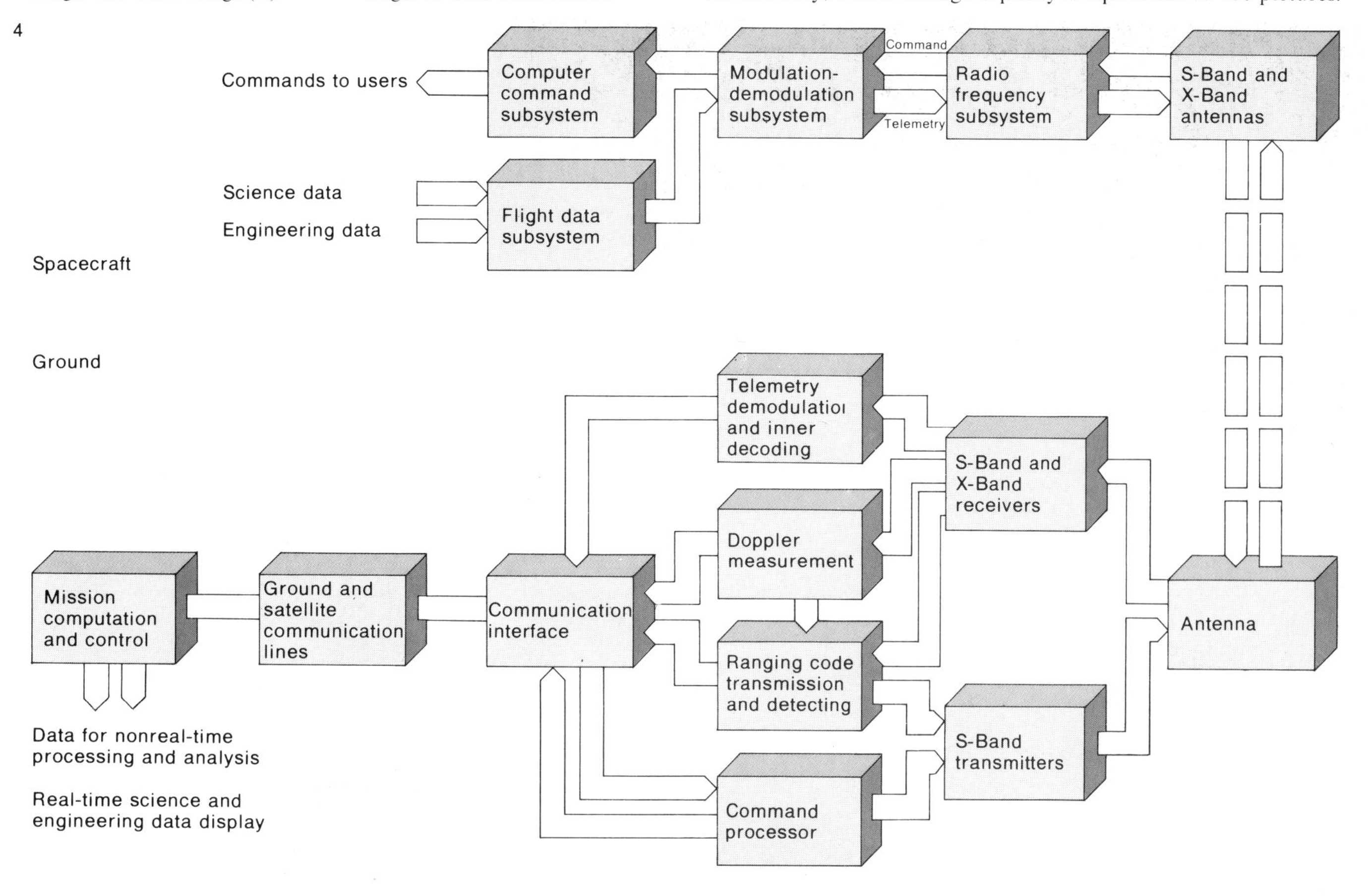

Structure of Saturn Interior

Planetary magnetism

There was no definite proof of a magnetic field associated with Saturn until Pioneer 11. It was reasonable to assume that a field should exist, since Jupiter was known to have a powerful magnetic field and in many ways Jupiter and Saturn are of a similar type. With Saturn, however, there is the added complication of the ring system, and it was thought that the rings might prevent energetic electrons from being trapped within a magnetic field.

The Pioneer and Voyager probes have, however, shown that there is indeed a strong magnetic field associated with Saturn. Generally speaking, the mechanism generating a magnetic field may be compared with a dynamo (*see* diagram 1). Some ordinary dynamos contain a permanent magnet, which provides the magnetic field, but it is also possible to construct a "self-exciting" dynamo, which generates its own field, and this probably illustrates the principle underlying the planetary magnetic fields studied so far. The two essential features are the presence of a good electrical conductor and a source of mechanical energy to drive the dynamo.

In the case of a planet, the electrical conductor is believed to be a conducting fluid at or near the planet's core. The force driving the system is still not known, and may involve a combination of several factors, but it certainly involves planetary rotation.

Interior

It is clear that the origin of Saturn's magnetic field must be closely related to the structure of the interior of the planet. The presence of a magnetic field, as in the case of Jupiter, may indeed be interpreted as evidence that parts of the interior are liquid, though it must be admitted that there is no direct evidence in either case. Theory indicates that the electrically conducting core of Saturn may be $0.46R_S$ (R_S standing for Saturn's radius), as against $0.7R_J$ (Jupiter's radius) for Jupiter.

Like Jupiter, Saturn consists primarily of the light elements hydrogen and helium, so that these giant worlds are quite unlike the terrestrial planets. The density of Saturn is only 0.7 times that of water: the terrestrial planets—Mercury, Venus, Mars and Earth—have mean densities in the region of five times the density of water. Both Saturn and Jupiter have gravitational fields that are powerful enough to retain the lightest elements, and models of their interiors necessarily depend heavily upon the assumed proportion of hydrogen to helium.

In this respect there is a marked difference between Jupiter and Saturn, since the Voyager observations show that there is much less helium, relatively speaking, in Saturn. Helium accounts for only about 11 percent of the mass of the atmosphere above the clouds of Saturn, as against 19 percent in the case of Jupiter. This indicates that the internal structures of the two planets are likely to be different from each other.

1. The principle of the dynamo
A simple model of a disc dynamo consists of a metal disc rotating in a magnetic field between two permanent magnets (**A**). The field produces a force on the free electrons in the disc, pushing them towards the center. As a result there is a difference in electrical potential between the edge and the center of the disc, which produces a current if the circuit is closed. In a self-exciting dynamo (**B**) this current is used to drive an electromagnetic coil, which replaces the original permanent magnets. The resulting system generates a magnetic field as long as the disc is kept spinning. The model thus demonstrates how mechanical energy may be converted into magnetic energy: an analogous process is thought to be responsible for creating planetary magnetic fields.

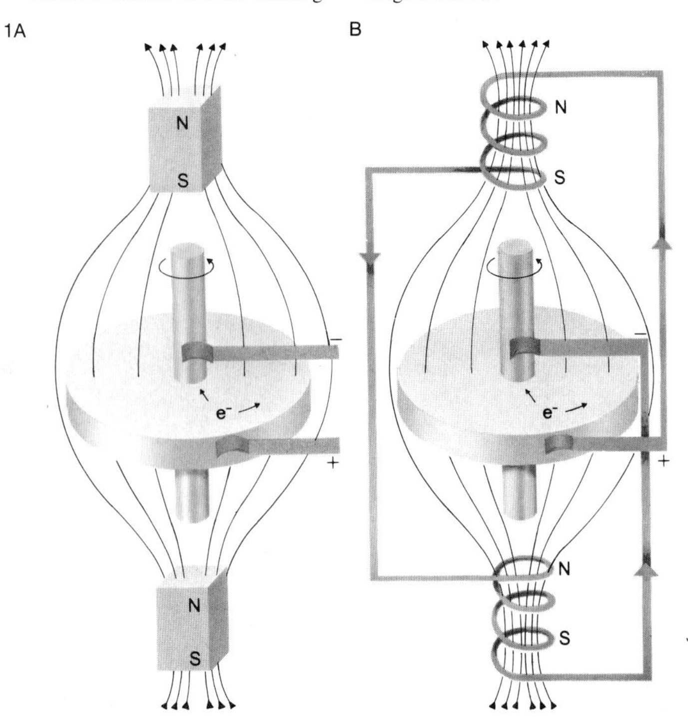

Internal structure

It is worth comparing the internal structures of Jupiter and Saturn according to the latest theoretical models. The transition from gaseous to liquid hydrogen in Jupiter takes place at a depth of about 1,000 km from the visible surface; the temperature at this level is about 2,000 K, and the pressure is 5,600 atmospheres (that is, 5,600 times the atmospheric pressure at sea level on the Earth's surface). At a depth of 3,000 km the temperature is 5,500 K and the pressure 90,000 atm, so that the hydrogen is highly compressed. At 25,000 km below the cloud tops—one third the radius of Jupiter—the temperature has risen to more than 11,000 K and the pressure to 3,000,000 atm. Under these conditions the hydrogen undergoes a dramatic change; it alters from its "liquid molecular" form to a state in which it is a good electrical conductor, and is then termed "liquid metallic hydrogen". Deeper still the temperatures and pressures rise, reaching about 30,000 K and 100 million atmospheres at the center. The core of Jupiter is rocky, with about 10 to 20 times the mass of the Earth; it is composed mainly of iron and silicate materials.

Saturn has a similar structure. A pressure of 1 atmosphere occurs just below the visible cloud tops, where the temperature is 140 K. The transition from molecular to metallic hydrogen occurs 32,000 km below the clouds, where the temperature is 9,000 K and there is a pressure of 3,000,000 atm. The outer boundary of the rocky, icy core has a temperature of 12,000 K, and the pressure is 8,000,000 atm. The core has a radius that is 16 percent of Saturn's total radius, so it is about the same size as the Earth, but three times as massive. The differences in the internal structures of Jupiter and Saturn may account for the different weather systems in the atmospheres of the two planets. Moreover, the presence of large quantities of liquid hydrogen is critical with regard to the existence, or otherwise, of a magnetic field.

Magnetic field

The existence of a Saturnian magnetic field was not confirmed until 1979. Saturn is so much more distant than Jupiter and it is not nearly as strong a radio source. The first definite proof came from Pioneer 11, which detected the presence of a magnetic field when it was still 1,440,000 km away from Saturn. There were further interesting results as the Pioneer passed beneath the outer edge of Ring A; the flux of charged particles was abruptly cut off, so that evidently there were well-marked interactions between the rings and the magnetic field.

Further results from the Voyagers have confirmed that the strength of Saturn's magnetic field is 1,000 times greater than the

Earth's, but about 20 times weaker than Jupiter's. At the cloud tops over Saturn's equator the strength is 0.22 Gauss: on Earth it is 0.3 Gauss at the surface. The magnetic axis is within one degree of the axis of rotation, in sharp contrast to the other known planetary magnetic fields, those of the Earth, Jupiter and Mercury, which are all tilted at about 10° with respect to the axis of rotation.

Saturn emits a radio pulse with a period of 10 hr 39.4 min, at frequencies of a few hundred hertz. This period may be regarded as the true rotation period of the planet, and is referred to as System III. The bursts of emission occur at local noon. Such a pulse would be easy enough to explain if Saturn's field were as complex and powerful as that of Jupiter, but it is not. The space-probes have provided few clues about the mechanism or mechanisms that are responsible for the pulse.

It is known, however, that the direction of polarization of the pulse is different in the two hemispheres of Saturn, and it therefore appears that the source is confined to narrow regions of latitude in the polar regions. It is worth noting that these polar latitudes—about 80° N and S—are also the regions where aurorae have been detected. Radio waves were detected from Jupiter in the 1950s, but Saturn is a much weaker source. It is smaller and has lower temperatures than Jupiter, and lacks a violently active satellite such as Io—the innermost Galilean satellite of Jupiter—which interacts extensively with its parent planet.

2

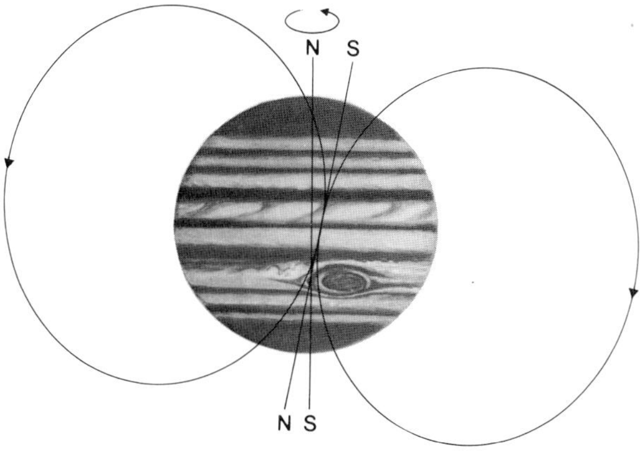

N S

N S

4

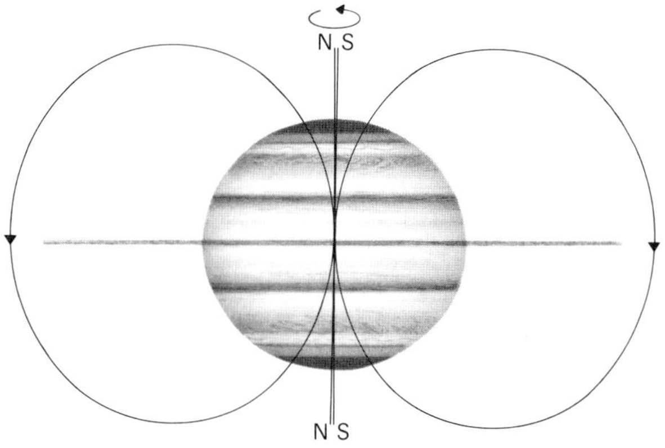

N S

N S

2. Jupiter's magnetic field
The magnetic field of Jupiter is inclined to the planet's axis of rotation by an angle of about 11°. Although the field is more complex than those of other planets, it still may be thought of as behaving as if a bar-magnet were embedded at the planet's core. Jupiter's magnetic field is 20 times greater than that of Saturn, which is 1,000 times greater than the magnetic field of the Earth.

3. Jupiter's interior
Jupiter is composed almost entirely of gases and liquids, although at present it is believed that there is a small core of rocky materials. The innermost part of the planet around the core is thought to consist of liquid metallic hydrogen, surrounded by a shell of liquid molecular hydrogen. The upper region consists of a deep atmospheric layer of hydrogen and helium.

4. Saturn's magnetic field
Saturn's magnetic field is stronger at the north pole (0.69 Gauss) than at the south pole (0.53 Gauss), the center of the field being displaced northward some 2,400 km along the axis of the planet. Saturn's field is unique in that its magnetic axis corresponds almost exactly to the axis of rotation: the angle between the two axes is less than one degree, much less than with other planets.

5. Saturn's interior
According to the theory put forward by W. B. Hubbard and his colleagues, Saturn has a rocky core the same size as Earth but three times as massive. This is overlain by a metallic hydrogen zone, which is smaller than Jupiter's because of Saturn's lower mass, gravitational field strength, and internal pressure. Above is molecular hydrogen, then a deep atmosphere.

3

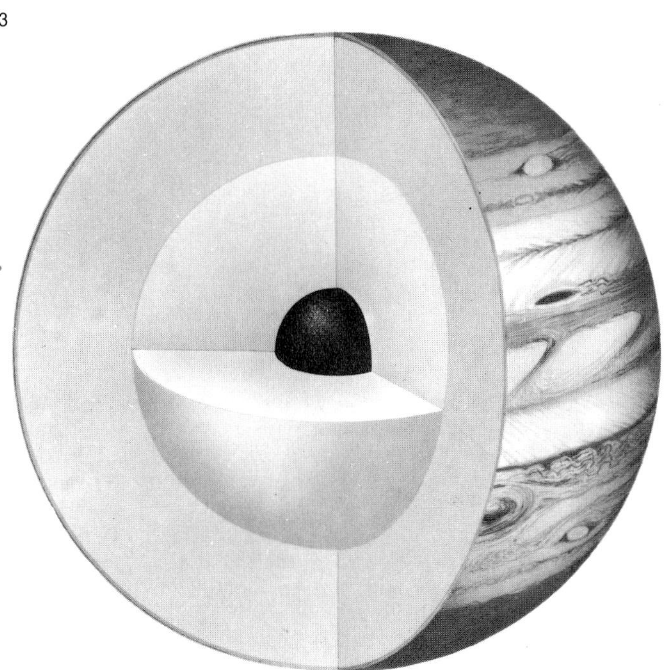

5

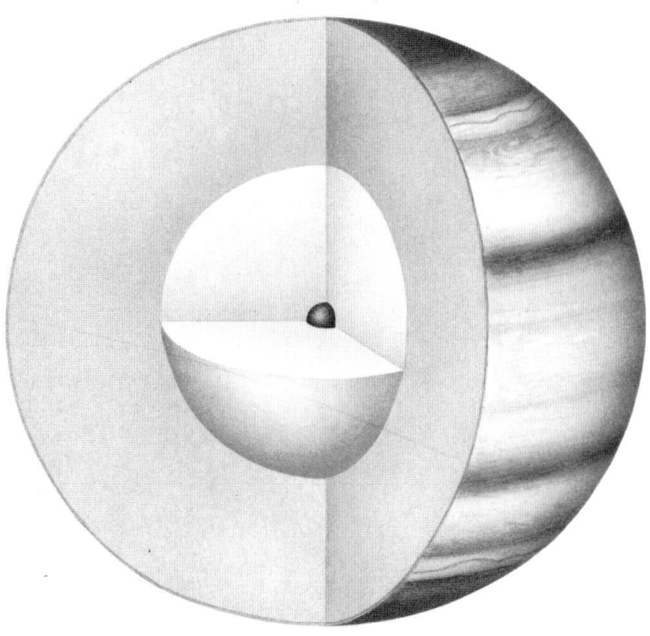

Structure of Saturn Magnetosphere

The magnetosphere of Saturn appears to be intermediate between those of the Earth and Jupiter, in terms of both extent and population of trapped energetic particles. The Earth's magnetosphere is of considerable extent, and Jupiter's even more so; indeed, the "magnetic tail" of Jupiter extends beyond the orbit of Saturn, so that at certain times Saturn lies within it. The presence of a magnetic field means that there must also be a magnetosphere, but that of Saturn was not definitely known before the flyby of Pioneer 11. It surrounds the planet like a giant bubble, extending away from the Sun like a wind-sock (*see* diagram 1).

Beyond the magnetosphere is the region dominated by the solar wind. This is made up of low-energy particles ejected from the solar atmosphere at speeds of up to 400 km/sec and in all directions; it is "gusty" in as much as its strength varies considerably, and this in turn affects planetary magnetospheres.

When the solar wind reaches the boundary of a planetary magnetosphere, it abruptly changes direction to avoid electromagnetic collision. The region in which this takes place is termed the bow shock; with Saturn, the average distance between the bow shock and the planet is 1,800,000 km. The magnetosphere itself lies considerably closer to Saturn, at a mean distance of about 500,000 km, but this distance is not constant, and changes according to the characteristics of the prevailing solar wind. Saturn's largest satellite, Titan—moving around the planet at a mean distance of 1,221,400 km, or $20.3R_S$—lies very close to the edge of the magnetosphere (*see* diagram 1), so that it is sometimes inside and sometimes outside. It was outside when Voyager 1 made its pass: the spacecraft made five crossings of the bow shock in all, at distances ranging between $26.1R_S$ and only $22.7R_S$. At the time of the Voyager 2 encounter, the magnetosphere was slightly more extensive, and Titan lay inside it, giving an opportunity for measurements of the interactions between Titan itself and the solar wind. Such interactions are the source of a huge, doughnut-shaped torus of hydrogen extending from the orbit of Titan inward as far as the orbit of the second-largest satellite, Rhea. Altogether Titan spends about 20 percent of the time inside Saturn's magnetosphere. The satellite is the source of both neutral and ionized molecules as a result of photodissociation in the upper atmosphere.

The magnetosphere is divided into several definite regions. Within a distance of about 400,000 km of Saturn there is a torus made up of ionized hydrogen and oxygen atoms. The plasma's ions and electrons spiral up and down magnetic field lines and contribute to the local field. At its outer edge some of these ions have been accelerated to high velocities. These indicate temperatures of from 400 million to 500 million degrees K.

Beyond the inner torus is a region of plasma that extends out to about 1,000,000 km, produced by material coming partly from Saturn's outer atmosphere and partly from Titan's. Titan is not

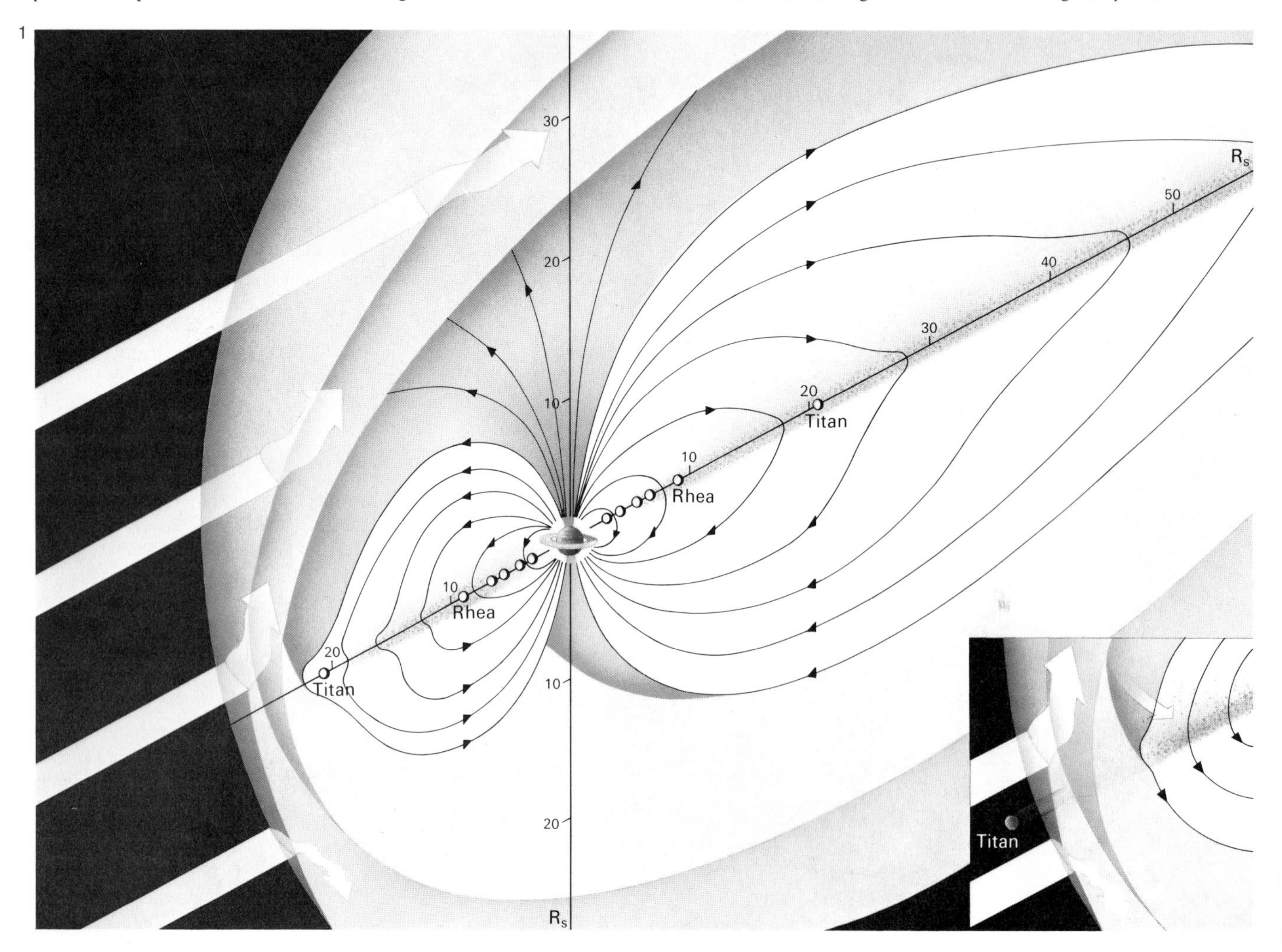

1. Saturn's magnetosphere
The region of space in which Saturn's magnetic field is dominant is less extensive than in the case of Jupiter. The region where the solar wind meets the magnetosphere is the bow shock, and this varies with the strength of the solar wind. The region immediately inside the bow shock is the magnetopause, and the whole magnetically active region is enveloped in a magnetosheath. Titan can lie inside or outside the magnetosphere (inset), thus affecting the satellite's interaction with the solar wind and the magnetosphere.

alone in having marked effects upon the magnetosphere; the other large inner satellites—Rhea, Dione, Tethys, Enceladus and Mimas—also play an important role (like their counterparts around Jupiter), although they are devoid of individual atmospheres. For example, protons in the magnetosphere are strongly absorbed by some of the inner satellites as well as the tenuous Ring E. The magnetic tail of Saturn has a diameter of $80R_s$ and is relatively devoid of plasma at high altitudes.

Of great importance are the interactions between the magnetosphere and Saturn's rings. This is, in fact, unique to Saturn since the much less spectacular rings of Jupiter and Uranus cannot have comparable effects. For example, the numbers of electrons fall off sharply at the outer edge of the exterior bright ring, Ring A, because the electrons are absorbed by the ring particles. It is also significant to note that the general magnetic field of Saturn excludes cosmic rays from the inner magnetosphere. (Cosmic rays are atomic nuclei coming from all directions in space, but they are electrically charged, and are influenced by magnetic fields.) Accordingly, the region between Ring A and Saturn is almost completely shielded, and is the most radiation-free region in the entire Solar System, excluding the atmospheres and solid globes of planets, asteroids, large satellites and the Sun.

Aurorae

Aurorae are common phenomena in the Earth's atmosphere: they occur as charged particles emitted by the Sun cascade into the upper atmosphere, producing the glows in the sky more commonly known as polar lights. The Voyagers had confirmed the presence of aurorae in the atmosphere of Jupiter, and subsequently found the same phenomena with Saturn, between latitudes 78° and 80° in each hemisphere. The presence of aurorae was not unexpected, although the displays had not been proved either from Pioneer 11 or from the artificial Earth satellite International Ultraviolet Explorer (IUE).

The brightness of aurorae is generally given in units termed Rayleighs; one Rayleigh is equal to 100,000 photons $cm^{-2}s^{-1}$. On Earth, an aurora with a brilliancy of 1,000 Rayleighs is just visible, and is an almost permanent feature of the polar night sky. The Jovian aurorae were found to be more energetic, reaching a value of 60,000 Rayleighs at ultraviolet wavelengths. The Saturnian aurorae attain from 2,000 to 5,000 Rayleighs, and seem to be associated with the positions of the edge of the "polar hoods" defined by lines of force of the magnetic field. Saturn's aurorae are weaker than Jupiter's by a factor of about 10.

Radio waves from Saturn

Saturn is not as rich a source of radio emissions as Jupiter, which is one reason why it was so difficult to establish the presence of a magnetosphere before the Pioneer and Voyager encounters. Yet at kilometer wavelengths Saturn is a powerful enough radio source; there is a broad band of emission extending from about 20 KHz to about 1 MHz. (One Hertz—abbreviation Hz—indicates a frequency of one cycle per second; Khz indicates a thousand hertz, and MHz a million hertz.) The maximum intensity occurs between 100 and 500 KHz, and there is a period of 10 hr 39.4 min, which is taken as the System III rotation period for Saturn.

The periodicity of the radio emissions is not easy to explain. A basic question concerning the radio emission is whether the rotational control is caused by a radiation pattern that rotates with the planet rather like a rotating searchlight. Alternatively, there could be a variation with time—like a flashing light. The apparent absence of a phase difference between the inbound and outbound Voyager observations implies that the radiation is emitted simultaneously over a wide range of directions in a particular plane of Saturn's rotation. Saturn differs from Jupiter, whose radio waves are strongly influenced by the inner, large and strongly volcanic Jovian satellite Io. It should be remembered also that the magnetic axis and the rotational axis of Saturn are almost exactly the same. Moreover, the situation changed between the passes of the two Voyagers. Voyager 2 detected the effects of Jupiter's magnetotail when approaching Saturn (*see* diagram 2), and soon afterwards, when Saturn was presumably immersed in the Jovian magnetotail, the kilometric radio emissions from Saturn were not detectable. It would, however, be premature to claim that this apparent shutting-off was due directly to the effects of Jupiter.

There are suggestions of a 2.7-day period in Saturn's radio emissions, and this is the period of one of the larger satellites, Dione, which may or may not be significant. However, Dione, unlike Io in Jupiter's system, is an inert, icy world. Other discrete, low-frequency radio emissions suggest that other satellites may be involved in the generation of radio emissions.

2

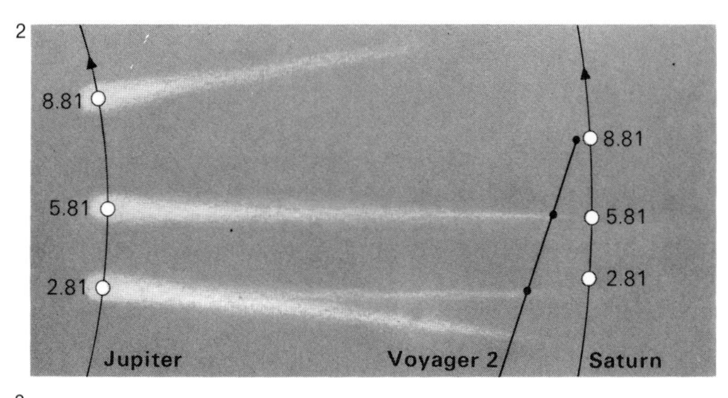

3

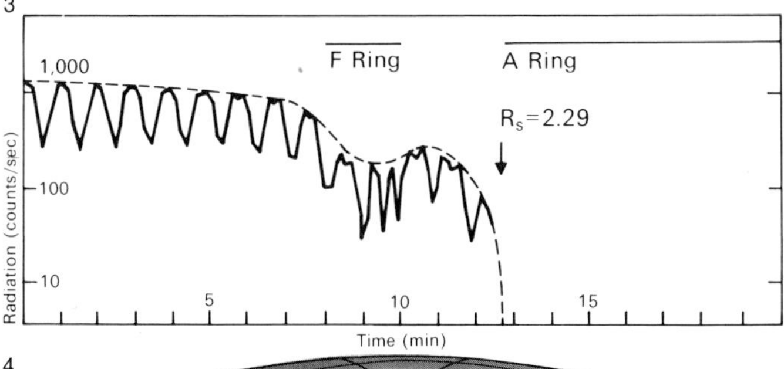

4

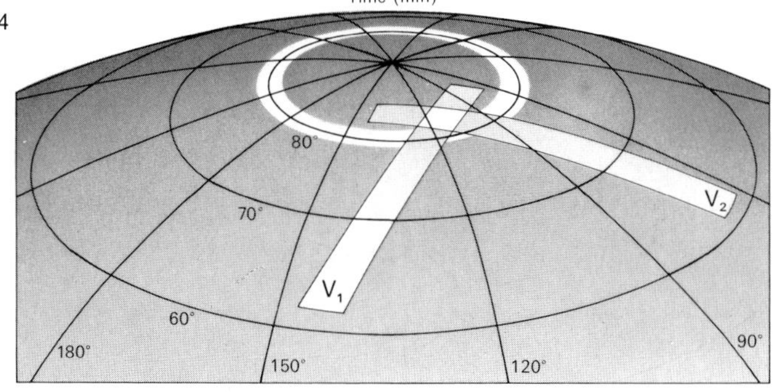

2. Jupiter's magnetotail
On its way to Saturn, Voyager 2 encountered the magnetotail of Jupiter in February 1981, even though it was not downstream of the planet until August. This may have been due to a branching filament or the rarefaction of the magnetotail, which was not detected at all in August.

3. Ring shielding
Radiation levels were recorded by Pioneer 11 as it approached Saturn. They show a dramatic reduction and then obliteration within the ring system. Electrically charged particles in space are affected by the magnetic field, which in this case acts as a shield.

4. Radio signals
Attempts to identify the source of Saturn's radio signals by both Voyager missions have narrowed the area down to a small, high-latitude region (where the two grey paths cross). Emission occurs only when this area crosses the noon meridian.

Structure of Saturn Atmosphere I

Thermal Radiation

When a body is heated, it radiates energy. If a curve is drawn, plotting the intensity of radiation against wavelength, it is found to have a characteristic form which reaches its peak at a certain wavelength depending upon the temperature of the body: the shorter the wavelength at which the peak occurs, the hotter the body. Radiation of this type is termed thermal, and is easily demonstrated in everyday life: for example, when an electric fire is switched on, the bars start to glow first dull red, then yellow, and eventually white or white tinged with blue. Physicists interpret this kind of thermal radiation in terms of what is called a "black body" (*see* Glossary). For such a black body it is possible to predict the exact intensity-wavelength curve for any given temperature.

Thus the thermal emission of Saturn, at wavelengths ranging from a few microns to centimeters, gives information about the temperatures at different levels in the planet's atmosphere. Radiation emitted from inside the planet passes through the atmosphere of Saturn before it is detected by instruments, and as it does so it is scattered by the cloud particles, so that it is absorbed and subsequently re-emitted. The molecules in the atmosphere absorb the radiation at different discrete wavelengths, so providing information about the chemical composition of Saturn's atmosphere. Extending the analysis beyond the thermal radio wavelengths into the infrared and visible range of the spectrum can also yield information about the variations of temperature and pressure at different levels in Saturn's atmosphere. If, for example, the temperature of the atmosphere increases with altitude gases at higher levels will appear to emit rather than absorb radiation.

Saturn's axis of rotation is inclined at 27° to the perpendicular to the orbit (as against 23°.5 for the Earth), and there are definite seasonal effects in the upper atmosphere during the long Saturnian year. The Sun crossed the equator into Saturn's northern hemisphere in early 1980, but the effects were not immediate; there is a prolonged lag, which explains the relative coldness of the northern hemisphere as recorded by the Voyagers.

Investigations of Saturn are more difficult than of Jupiter, because there are no wavelengths at which the clouds are sufficiently transparent for the lower levels of the atmosphere to be probed. It is possible, however, to produce maps of the thermal radiation at different wavelengths.

At the time of Voyager 2's encounter the temperature was about 10 K lower in the north polar region than in the south. Strong latitudinal temperature gradients were found in the troposphere of the northern hemisphere. These were related to strong westerly jets. The largest longitudinal variations were found in the regions between 32° and 42°N, 50° and 57°N, and between 65°N and the pole. The first two regions indicate relatively weak retrograde jets.

The temperature structure in the troposhere of Saturn is symmetrical with respect to the equator—a surprising result in view of the axial inclination, which would be expected to produce seasonal effects. The probable explanation is to be found in the low level of solar energy in these remote parts of the Solar System: the available energy is only about one hundredth of that received upon the Earth, and Saturn's atmosphere has a very slow response to radiation changes. Perhaps an even more surprising discovery is that there is little connection between Saturn's temperature structure and the visible cloud belts.

Heat budget

The total energy budget of a planetary atmosphere is an important parameter relating to the mechanisms of weather systems. Saturn, like Jupiter, has a strong internal heat source. Jupiter radiates about 1.6 times as much energy as it would do if it depended entirely upon what it receives from the Sun. The value for Saturn is also about 1.6. Both Jupiter and Saturn were formed about 4,600 million years ago, mainly from hydrogen and helium. If it is

Composition of Jupiter's atmosphere above cloud tops

	% volume
Hydrogen	≈90
HD	≈ 1.8×10^{-3}
Helium	≈4.5
Methane	≈ 7×10^{-2}
Deuterated methane	≈ 3×10^{-5}
Ammonia	≈ 2×10^{-2}
Ethane	≈ 10^{-2}
Acetylene	≈ 10^{-2}
Water vapor	≈ 1×10^{-4}
Phosphine	≈ 10^{-6}
Carbon monoxide	≈ 10^{-7}
Germanium tetrahydride	≈ 10^{-7}

Composition of Saturn's atmosphere

	% volume
Hydrogen	≈94
Helium	≈6
Ammonia	≈ 2×10^{-4}
Phosphine	≈ 1×10^{-6}
Methane	≈ 8×10^{-4}
Ethane	≈ 5×10^{-6}
Acetylene	≈ 2×10^{-8}
Methylacetylene	≈ 10^{-10}
Propane	≈ 10^{-10}
HD	≈ 5×10^{-5}
CH_3D	≈ 2×10^{-5}

1

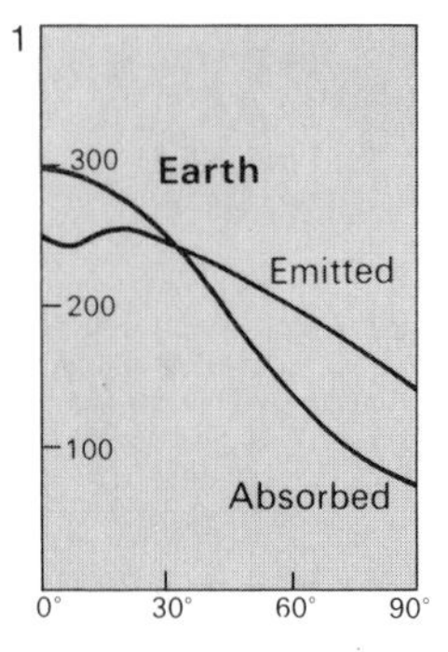

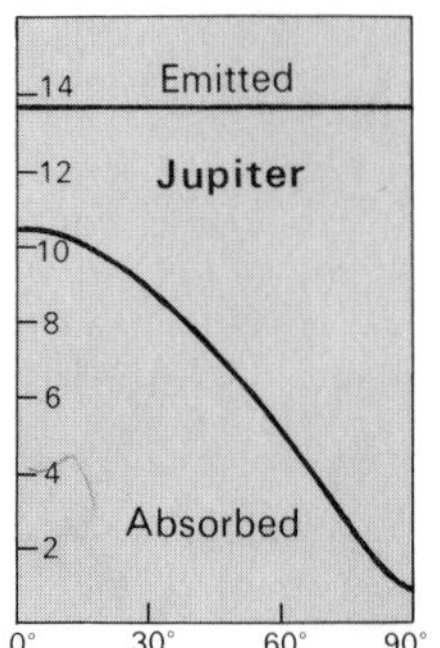

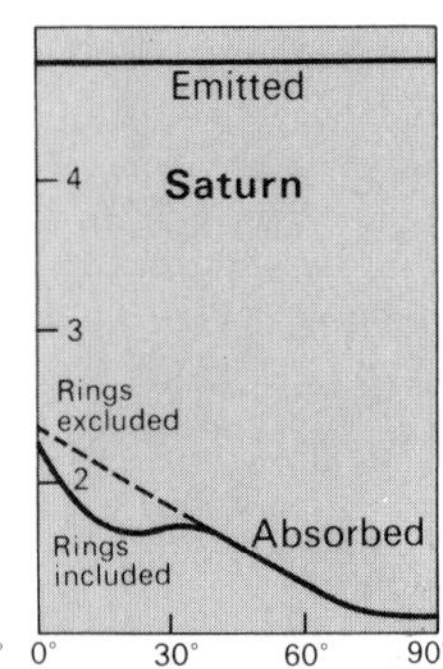

2

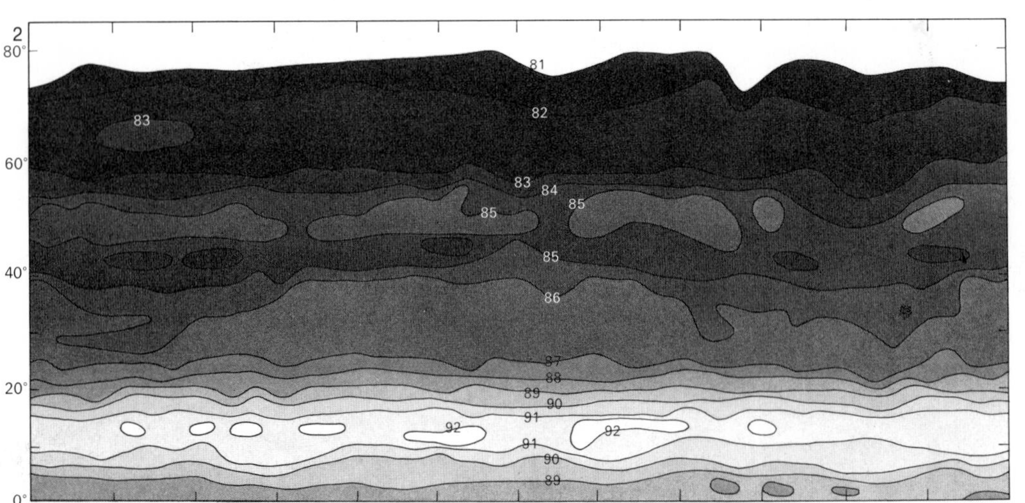

1. Heat budget
A comparison of absorbed solar energy and emitted infrared radiation has been plotted for the Earth, Jupiter and Saturn. Values have been averaged for season, time of day and longitude. Both Jupiter and Saturn emit more infrared than sunlight absorbed, and the constancy of emission suggests heat transport across latitude circles deep within the planets. Saturn's seasonality and rings make figures uncertain.

2. Thermal map
A map of brightness temperatures within the spectral interval 330 to 400 cm^{-1}, corresponding to a region with 150 mb pressure, was plotted during Voyager 2. The tropopause was 10 K cooler in the northern hemisphere (winter).

assumed simply that their globes have cooled down since then, the computed value for Jupiter's excess radiation can be accounted for, but with Saturn things are less straightforward: there is more emitted energy than there ought to be.

Since Saturn is much less massive than Jupiter, it has cooled faster, and its present internal heat is less than Jupiter's. About 2,000 million years ago the internal temperature dropped below the critical point where helium condensed on the surface of the fluid core of metallic hydrogen. At this point droplets of helium formed, and fell in what might be described as rain, dropping through the hydrogen fluid and releasing heat by their interactions with it. This seems to account for the discrepancy. In the case of Jupiter, the critical point at which helium condenses is only now being reached in the planet's interior. Extra proof is provided by the fact that there is relatively less helium in Saturn's atmosphere than in Jupiter's.

The internal heat sources in Jupiter and Saturn, in addition to weak solar heating, drive meteorological systems. The internal heating is not likely to be uniform with latitude, as with the terrestrial planets, and, in effect, only differential solar heating operates; with Jupiter and Saturn the net heat loss is greatest at the poles and least at the equator.

The structure of the atmosphere

The atmospheres of planets are conventionally divided into certain well-defined regions. The lowest is the troposphere. On Earth the temperature above the surface decreases with altitude until it reaches a height of about 15 km. One of the main constituents of the troposphere is water vapor and it is there that weather processes, in particular the formation of clouds, take place. The equivalent region on Saturn is bounded at the upper level by a temperature minimum of about 90 K and a pressure of 100 millibars (mb) (compared with 1,000 mb at sea-level on Earth).

Above the troposphere is a transition region, the tropopause, above which is the stratosphere. The temperature in the Earth's stratosphere is controlled largely by radiation processes and is relatively high. This temperature inversion is the result of the heating of the upper atmosphere of the ozone layer, which absorbs ultraviolet radiation from the Sun. The same effect occurs on Saturn, but here the cause is to be found in methane.

On Earth, the temperature gradient changes again at an altitude of about 30 km, decreasing with height for a further 30 to 40 km. This is the mesosphere. It is assumed that the same applies to Saturn, but there is not yet enough information to confirm this.

The ionosphere

Above the mesosphere is the ionosphere, where the density is very low indeed, and the electrical conductivity increases. The name is derived from the higher proportion of atoms and molecules which are ionized, that is stripped of one or more of their electrons: the increased conductivity is due to the consequent presence of greater numbers of free electrons.

The Voyager 1 observations, made at wavelengths of 3.6 and 12 cm, provided the first look at Saturn's ionosphere at high latitudes, around 73°S. The atmosphere in this region is highly rarefied, and the molecules are easily ionized by energetic radiation from the Sun at wavelengths in the ultraviolet region less than 1,000 Ångströms. Energy from the Sun in the ultraviolet part of the spectrum is absorbed by the atoms and molecules in the ionosphere, causing them to lose electrons and become positively charged ions.

The extreme ultraviolet solar radiation which causes ionization originates in the Sun's upper atmosphere and corona. Changes in these parts of the solar atmosphere therefore produce detectable variations of density in the ionosphere. The main constituent of Saturn's ionosphere is ionized hydrogen (H^+ ions) as with the Earth and Jupiter. The structure of the ionosphere varies according to the degree of solar activity. The ionosphere of Saturn is modified by a wider range of physical effects than any other similar structure yet studied in the Solar System.

The composition of the atmosphere

Hydrogen is the most abundant element (about 94 percent) in Saturn's atmosphere, then helium (about 6 percent). The main difference between the atmospheres of Jupiter and Saturn is the lesser quantity of helium in the case of Saturn. Detailed comparisons of the spectra of Jupiter and Saturn reveal other important differences. In the troposphere, ammonia is much less prominent on Saturn because of the lower temperature, which makes the ammonia condense out at higher levels. Phosphine (PH_3) is dominant in the region of the spectrum centered on 1,000 cm^{-1}. On Jupiter it has been found that phosphine demonstrates strong convective processes, bringing it up from lower levels, where the temperature is about 2,000 K, to a region where the temperature is only about 100 K. It is thought that vertical motions on Saturn are more violent than those of Jupiter (*see* pages 24–25).

In the stratosphere, emissions due to CH_4 and its photochemical derivatives, acetylene (C_2H_2) and ethane (C_2H_6), are seen on both planets. The concentrations of these two gases are approximately the same in each case, although there are minor variations due to differences in the pressures and temperatures involved. Traces of propane (C_3H_8) and methylacetylene (C_3H_4) have been detected in Saturn's atmosphere.

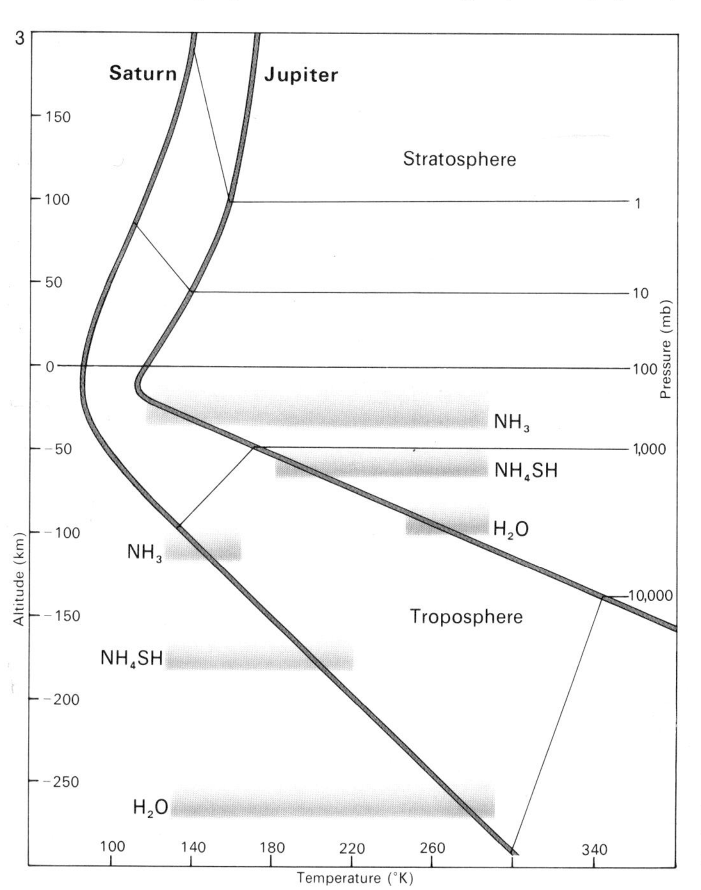

3. Atmosphere profiles
Profiles of temperature and pressure in the upper atmospheres of Jupiter and Saturn are based on measurements at infrared and radio wavelengths. The colored bands indicate the altitudes where various cloud layers should form—based on a solar composition of gases. The altitude is shown in kilometers above the 100 mb mark. Temperatures are lower on Saturn because of its greater distance from the Sun, and the atmosphere is more distended than that of Jupiter as a result of Saturn's weaker gravity, so the cloud layers are broader.

Structure of Saturn Atmosphere II

One of the basic differences between the visible appearances of Jupiter and Saturn is the apparent scarcity of distinct cloud systems in the atmosphere of Saturn. Features with diameters greater than 1,000 km are at least ten times rarer than on Jupiter. The largest feature observed by the Voyagers is about the size of a Jovian white oval and only about half the size of Jupiter's Great Red Spot.

The lower contrast of features on Saturn is explained in part by lower temperatures and weaker gravity. Since Saturn's atmosphere is colder than Jupiter's, the condensation point of compounds such as ammonia is reached at a higher pressure (greater depth) on Saturn. The pressure at the 150 K level, for example, is at 0.7 bar on Jupiter, but as much as 1.4 bars on Saturn, while the tropopause is at 0.1 bar (that is, 100 millibars) on each planet. Taking into account the difference in gravity, the mass per unit area between the 150 K level and the tropopause is about $2 \times 10^3 \text{kgm}^{-2}$ for Jupiter and about 10^4kgm^{-2} for Saturn. If a haze of ammonia or other particles was mixed throughout these layers in the same ratios relative to gas, the mass per unit area would be five times greater for Saturn. As a consequence, the colors and contrasts on Saturn would be expected to be greatly reduced in comparison with Jupiter, and this is confirmed by observations. These extensive clouds and hazes may well mask Saturn's dynamic cloud activity.

Circulation of the atmosphere

The meteorological phenomena that produce the visible cloud structures take place in the troposphere. Saturn, like Jupiter, is an essentially liquid planet with only a very small solid core, so that unlike the Earth, Mars and Venus there is no solid surface to produce observable effects. The driving mechanisms in Saturn's atmosphere are also essentially different from those of the terrestrial planets. For example, there is only a small temperature gradient between the poles and the equator: the difference is a mere 5 K or so, and the same is true of Jupiter, where polar and equatorial temperatures are very much the same. On Earth, the poles are much colder than the equator, and so the transfer of heat from the equator plays an important role in terrestrial meteorology.

A further important difference between Jupiter and Saturn on the one hand and the Earth on the other, is the heat source that drives the weather systems. On Earth, the principal source of energy is the radiation received from the Sun, but both Jupiter and Saturn have additional internal energy, which affects the circulation. Moreover, the fact that both Jupiter and Saturn are rapidly rotating planets also constrains the motions.

Nevertheless, many of the individual features on Saturn's visible surface may be interpreted by analogies with terrestrial weather systems. For example, on Earth air will flow outward from a high-pressure region, producing an anticyclone, which spirals in a clockwise direction in the northern hemisphere and anticlockwise in the southern hemisphere. With low-pressure systems the direction of flow is inward, producing a cyclone, which spirals in a direction opposite to that of an anticyclone in each hemisphere.

The first detailed measurements of the wind patterns of Saturn's clouds, made from the analyses of Voyager images, showed characteristics that differed markedly from those on Jupiter. On Saturn, all the velocities are relative to System III, which is the solid-body rotational period. The broad equatorial jet, with a peak velocity of more than 500 ms^{-1}, moves at approximately two thirds of the speed of sound in the region where the cloud temperature is about 100 K. The westerly flow spreads over more than 35 degrees in each hemisphere. There are only three easterly jets, symmetrically placed in each hemisphere at latitudes of about 40, 58 and 70 degrees respectively. As with Jupiter, these easterly jets mark regions of unstable flow, and, moreover, the overall flow pattern seems to be stable over long periods.

Saturn also differs from Jupiter in that the easterly and westerly flow patterns of zonal winds do not closely correlate with the light and dark cloud bands. A further complication arises from the variations in the boundaries of these bands when Saturn is observed at different wavelengths ranging between ultraviolet and red in the solar spectrum, while the zonal velocity profile remains unaltered.

1. Zonal velocities
The graph shows the relative zonal velocities of Saturn's mean eastward winds at various latitudes. They are measured relative to the planet's rotation and are plotted alongside an image showing the bands. Negative velocities are winds moving westward. There is a symmetry between northern and southern hemispheres.

1

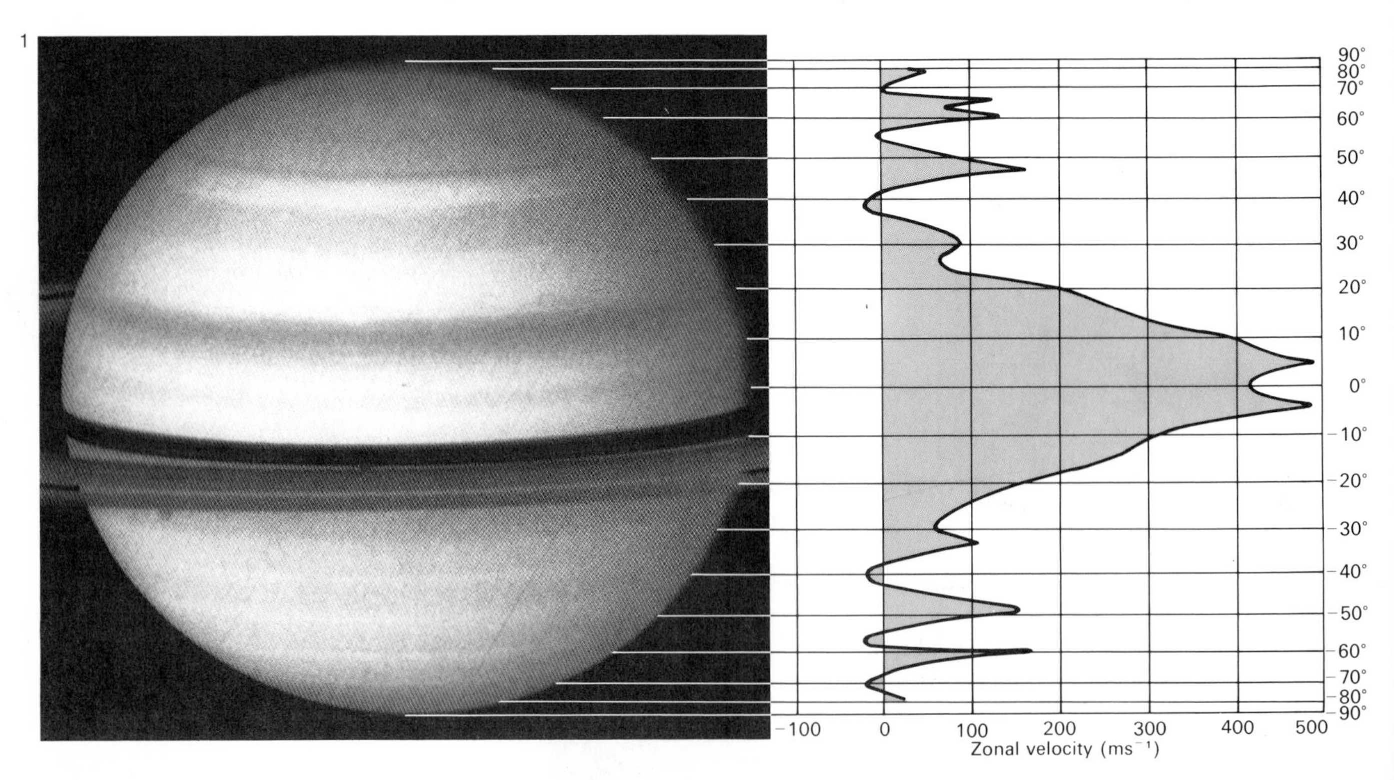

This suggests that the cloud movements observed are simply the average for a vertical layer whose thickness is less than 20 km.

A particularly surprising observation is that the Saturnian winds are completely symmetrical with respect to the equator (*see* diagram 1). This seems to be in complete contrast with the 27-degree inclination of the axis of rotation, and indicates that the weak radiation received from the Sun has only a slight effect upon the observed motions of the clouds. This raises the basic question of how deeply the motions extend into the atmosphere.

One possible explanation may be related to the interior structure of Saturn: the zonal velocity profile may extend, essentially unchanged, deep into the planet. If the liquid interiors of Jupiter and Saturn are adiabatic (without transference of heat) any steady zonal motion will take the form of differentially rotating, concentric cylinders (*see* diagram 2) whose common axis coincides with the planet's axis of rotation. Cylinders that do not reach the metallic core will extend from the top of the adiabatic zone in the north to the corresponding top of the adiabatic zone in the south. The zonal velocity profiles would, therefore, be symmetrical from north to south at the top of this zone, which may well extend to the base of the clouds. Any departures from symmetry would then arise as a result of non-adiabatic conditions within this cloud zone.

If this is true, then the density change at the top of the metallic hydrogen layer will decouple Saturn's northern and southern hemispheres poleward of about latitude 65°. For Jupiter the decoupling will take place at latitude 40° to 45°, because the metallic hydrogen core is larger. In each hemisphere of both planets, there are about three complete cycles of alternating easterly and westerly winds from the equator to the latitudes at which decoupling occurs.

This "deep-atmosphere" model provides a satisfactory explanation of the zonal motions, but the energy budget of the motions must be taken into account. With Jupiter, it has been found that the motions result from energy being transported from eddies into the mean zonal flow. A similar process helps to maintain the Earth's jet-streams, but there is an important difference. On Earth, this energy transfer, averaged over the planet, is only about one thousandth of the total energy flowing through the atmosphere as sunlight and infrared radiation. On Jupiter, the transfer of energy by this mechanism is more than one tenth of the energy flow, so that Jupiter is able to harness the thermal energy flow a hundred times more efficiently than the Earth's atmosphere. This may be due to the fact that on Earth the mid-latitudinal cyclones and anticyclones obtain their energy from a horizontal transfer of heat due to the pole–equator temperature difference, while the Jovian eddies obtain half their energy from convection from the interior.

Saturn is likely to behave in a similar way, but since the interior heat provides more than twice the energy from solar heating, the buoyancy is presumably much greater, thereby accounting for the much stronger vertical motion in the clouds of Saturn as compared with those of Jupiter.

In recent years considerable progress has been made in studies of the Earth's meteorology, using numerical models which give detailed representations of the physical processes and surface conditions over the planet. One such model has been extended to the conditions on Jupiter and Saturn (*see* diagram 3). With this type of approach it is possible to vary different parameters so as to examine the response of the atmosphere. The results, generated by a computer, closely resemble the large-scale frequencies of the atmospheres of Jupiter and Saturn, notably the main easterly and westerly jet profiles. It would seem that instabilities in Saturn's atmosphere provide the energy responsible for the overall circulation, creating large-scale waves resembling the belts and zones, while the jets are due to the interactions of propagating waves.

2. Cylinder model
This model shows the large-scale flow possible within the molecular fluid envelope of Saturn. There is a unique rotation rate for each cylinder and the zonal winds (diagram 1) may be the surface manifestation of this phenomenon. The tendency of fluids in a rotating body to align with the axis of rotation is likely if Saturn's interior is adiabatic.

3. Computer model
This model shows the possible atmospheric circulation on Saturn, assuming that the energy to power such circulation comes from within a narrow atmospheric layer (as on Earth); vertical heat transport is not accounted for. In the model small-scale eddies (**A**) become unstable (**B**), producing zonal jets (**C,D**). These eventually dominate the circulation (**E,F**).

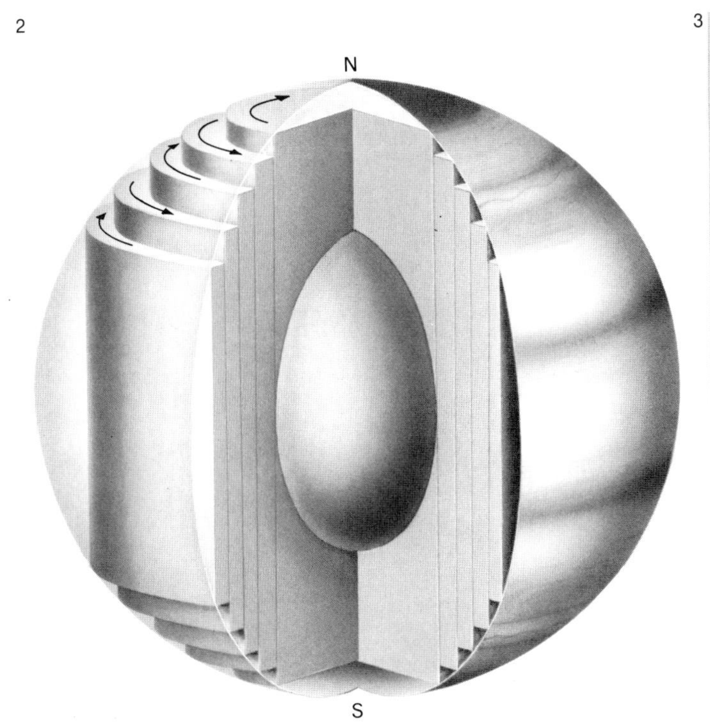

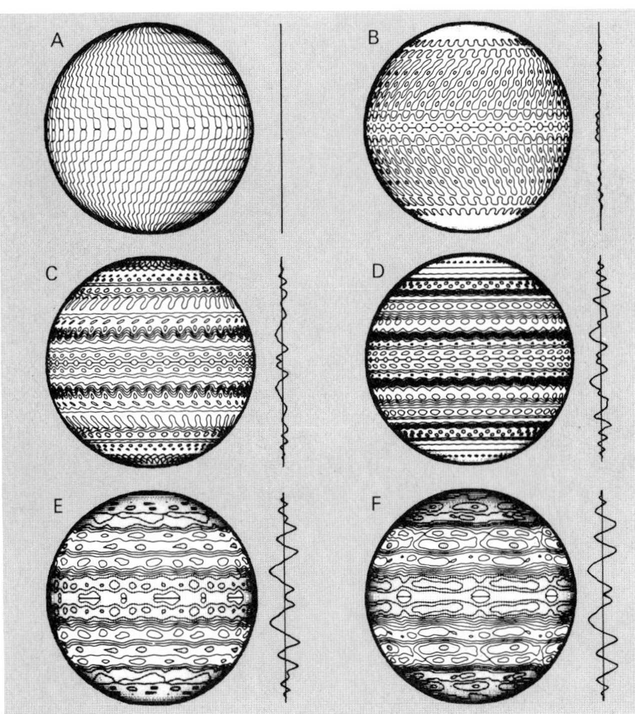

Structure of Saturn Cloud Morphology I

Saturn's greater distance from the Sun as compared with Jupiter results in lower light levels, which, together with the lower contrast and less conspicuous features of the disc, make images of Saturn much more difficult. Voyager 2 images, however, greatly extended our information about the dynamics of the Saturnian atmosphere and provided many more details of the cloud systems.

The greater detail revealed many similarities with the cloud features of Jupiter. Long-lived oval spots and tilted features in east–west shear zones are similar characteristics, as are high-speed jet streams, alternating between eastward and westward directions with increasing latitude. The jet streams are thought to be powered by small-scale eddies as in the Jovian atmosphere. Greater wind speeds and latitudinal spacing of the zonal jets on Saturn are the major differences. Winds are up to four times stronger than on Jupiter. The zonal jets are twice or even four times as wide and bear little relation to the banded cloud structure.

The pattern of jet streams on Saturn occurs at higher latitudes. There is a dominance of eastward jet streams which tends to suggest that the winds are not restricted to the cloud layer but extend down by at least 2,000 km into the atmosphere. They may extend even further. The symmetrical pattern of jet streams in the northern and southern hemispheres is consistent with the cylindrical model of the structure of the atmosphere (*see* page 25).

The extensive rings of Saturn cast a huge shadow on the planet's clouds. Although the rings obscure sunlight from regions near the equator, this does not have any noticeable effect on planetary weather systems. Saturn's atmosphere behaves rather like a huge ocean, and therefore responds very slowly to the weak solar radiation, and over a length of time that is greater than that of the passing ring shadow.

Northern hemisphere

In the region extending to 7°N, wispy cloud structures have been observed which move very rapidly, the cloud-top winds attaining velocities of up to 500 ms^{-1}. As with Jupiter, there is a minimum zonal velocity at the equator. Wisps of cloud in this region of Saturn's atmosphere are inclined in a way that tends to support this conclusion.

As in the case of Jupiter, there is a wide range of atmospheric cloud systems on Saturn. Several stable symmetric ovals of various colors (white, brown and red) have been observed at several latitudes. Most of the features are located in the anticylonic shear zones. A shearing action is set up by a decrease in wind velocity with increasing latitude. At 27°N, a feature prominent at ultraviolet wavelengths (the UV Spot), and, therefore, higher than the surrounding clouds, was observed throughout both Voyager encounters. Three brown spots are situated at 42°N. Brown spot 1 is 5,000 by 3,300 km. The flow characteristics of the Saturnian brown spots are similar to those of the Jovian white ovals. The

1

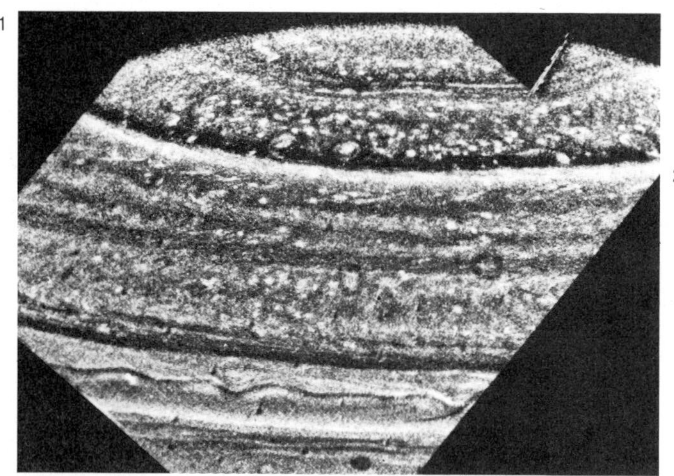

1. Northern hemisphere
This Voyager 2 image of the northern hemisphere shows how, at close quarters, the light and dark banding breaks down into a variety of very different smaller features. Several cloud shapes suggest motion, some indicating convection currents from lower levels in Saturn's atmosphere. The ribbon-like feature is as yet unique to this planet.

2

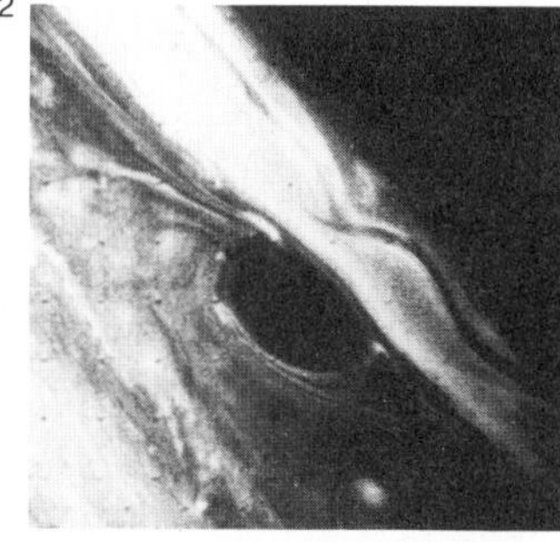

3

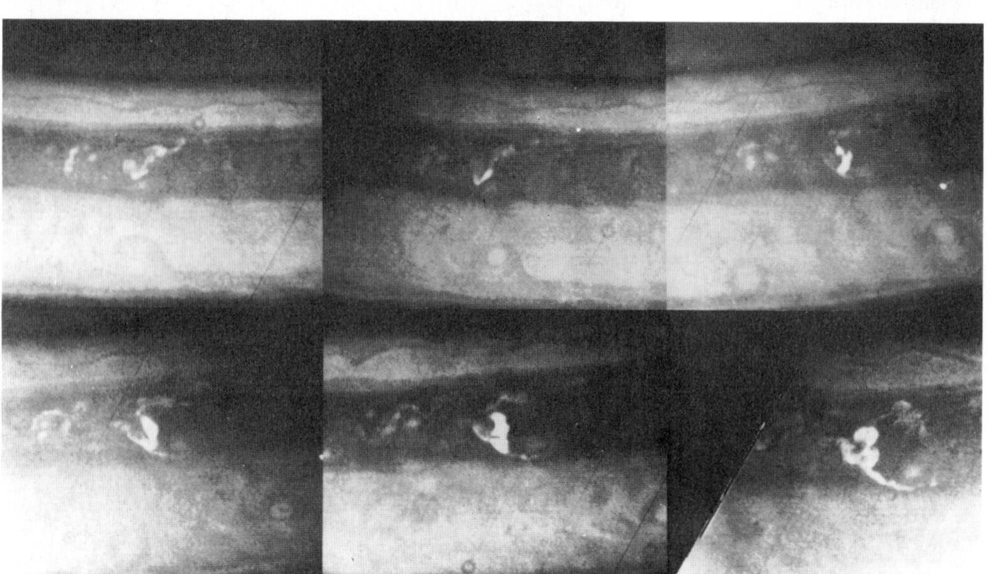

2. Brown Spot
The evolution of Brown Spot 1 during two successive rotations of Saturn shows evidence of anticyclonic rotation (clockwise direction in the northern hemisphere) around the edge of the spot. This feature is at latitude 42°.5N and measures some 5,000 km across. These images were taken about two days before Voyager 2's encounter.

3. Convective features
There are intermittent eruptions of convective clouds similar to the equatorial plumes on Jupiter. They flow with a westward jet at 39°N and appear to originate in several source regions in this latitude band. Why they are restricted to this region is not known. Individual cloud components of the feature are bright, white, irregular in shape and short-lived. To the south (27°N) is the UV Spot.

largest stable oval, nicknamed Big Bertha, is a reddish cloud measuring 10,000 by 6,000 km and is situated at 72°N.

In the northern polar regions two hurricane cloud systems have been observed at about 72°N. Each is 250 km in diameter with a distinctive core 60 km in diameter. The overall structure and size of these systems is very similar to those of terrestrial hurricane systems. On Saturn, these convective cloud systems are thought to be driven primarily by energy released by condensation processes in the water ice layer beneath the visible ammonia clouds. The strong upward motions are also influenced by the internal heat source of the planet itself. At the highest level, ammonia rains out, so that condensates are recycled, thus prolonging the lifetime of the cloud disturbance. On Earth, however, such storms are strongly influenced by the availability of a warm, moist surface layer.

Convective features associated with the strong easterly jet at 39°N were prominent in the Voyager 2 observations, and resembled the equatorial plumes and disturbances at higher latitudes in the Jovian atmosphere. Vertical motions associated with these features are much stronger than the Jovian plumes. This is a region which altered its appearance significantly during the nine months between the Voyager encounters. The extensive cloud systems situated between 20° and 35°N during the first Voyager encounter dispersed during the following months, revealing a great many previously unknown cloud details. One feature first appeared as a "cloud knot" with a diameter of 3,000 km. Within 64 hr the shearing action of the jet streams coiled the cloud into a tight oval resembling a figure 6, before the cloud ultimately detached itself from the jet to the north. The cloud oval appeared to have a total lifetime of about two weeks.

Numerous unstable cloud storms seem to originate in this region, where the wind speeds are at a minimum. One system resembles a train of vortices that might be shed by a solid vertical cylinder moving to the west. In the laboratory the characteristics of a flow of liquid are affected by the presence of a solid cylinder in the flow, an effect called a "Taylor column". This type of situation can also be seen on Earth with the air flow over the Canary Islands.

On Saturn a deep convective tower originating in the hotter lower levels of the globe acts as a barrier to the flow, since it is situated in a region of at most a very weak westerly motion. The change in the convective processes causes the tower to pulsate and spawn smaller eddies to the north: these eddies become entrained in adjacent westerly moving jets. This process could provide an almost continuous source of energy for the zonal winds.

The ribbon-like feature at 46°N latitude may be unique to Saturn. This dark, wavy line moves with the peak westerly winds of about 150 ms^{-1} at this latitude. Each crest or trough covers about 5,000 km in an east–west direction. To the north of the ribbon, cyclonic vortices nestle in the troughs, and their filaments spiral towards the center in an anticlockwise direction. To the south, under the crests, are anticylonic vortices which spiral inward in a clockwise direction. A further ribbon-like feature appears at 78°N on Saturn, with a length of 16,000 km; this coincides with the jet further to the northwest.

A complex interaction between two easterly white spots moving at 15 and 20 ms^{-1} respectively, and a brown spot moving at 5 ms^{-1} in a westerly direction, was observed (*see* illustration 4). In the beginning the two white spots were at approximately the same latitude, and about 10,000 and 15,000 km to the east of the brown spot. All the spots exhibited an anticyclonic circulation. At the beginning of the sequence the furthest white spot (WS1) appeared to be slightly south of the second feature (WS2) and, if they had followed the mean flow, they would have moved closer together as WS2 overtook WS1. However, instead of colliding, WS1 moved further south and went around WS2.

Four days later WS2 passed WS1. While some merging did take place, there appeared to be a connected circulation between the two spots. The dark band stretched out considerably during the next few days as WS2 continued to move in a westerly direction relative to WS1. After six days the band stretched more than 500 km and during the next two rotations of Saturn the dark band became much narrower, while the brown spot approached to within a few thousand kilometers of the two white spots. The band between these spots appeared to be on the verge of disappearing a day later. The interaction between the three spots was a new phenomenon in atmospheric fluid dynamics.

4. Interaction of spots
The mutual action of two white spots as they approached Brown Spot 1 was tracked by Voyager 2 over the course of seven days. The first image (**A**) was taken at 8.3 million km from Voyager's closest approach to Saturn. Frame **B** was taken 3 Saturn rotations later and images **C** to **G** followed at intervals of 4, 2, 5, 1 and 1 rotations respectively. By image **G** the Voyager was only 1.3 million km from the planet. The width of each panel is approximately 16,000 km.

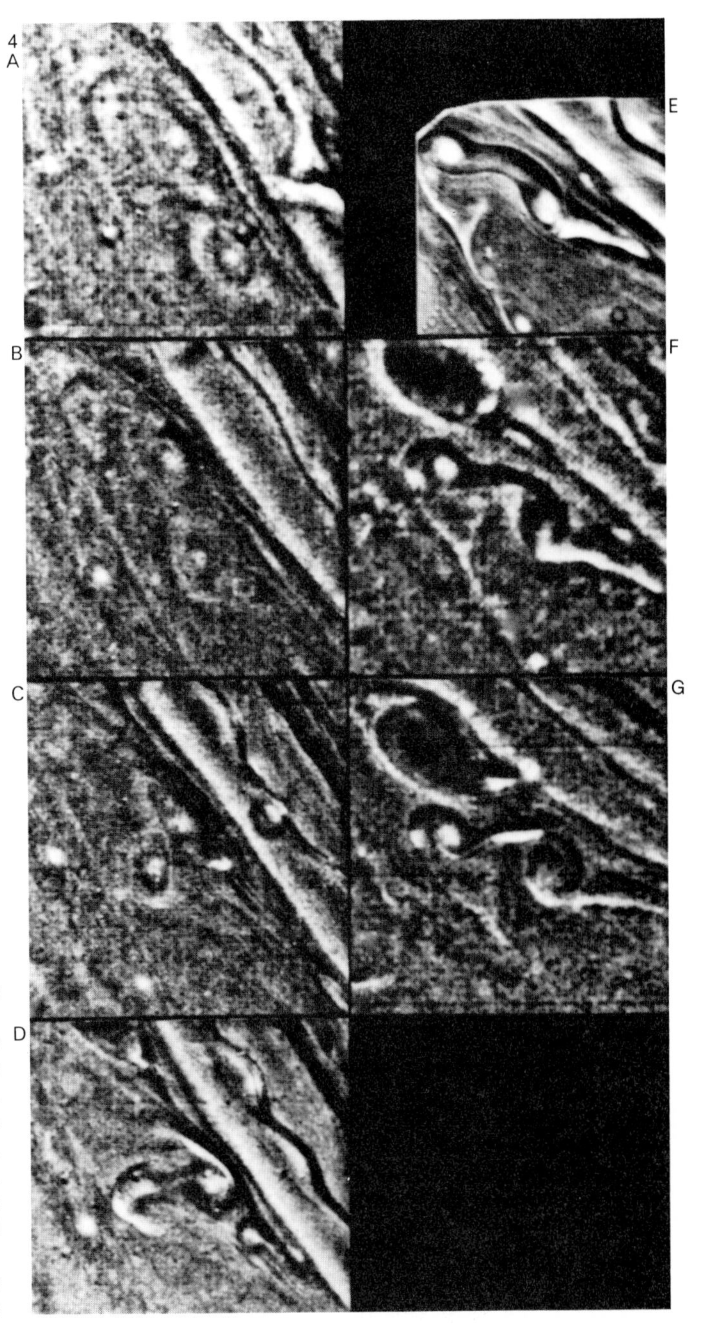

Structure of Saturn Cloud Morphology II

Southern hemisphere

The major feature observed in the southern hemisphere of Saturn is the large red spot (Anne's Spot) situated at 55°S, which measures about 5,000 by 3,000 km. The feature was first observed in August 1980, and was tracked throughout both Voyager encounters until September 1981. It resides in a region where there is a westerly flow at about 20 ms^{-1}. The red color of the feature is thought to be the result of the same processes as those operating within the Great Red Spot on Jupiter. Saturn has a large amount of phosphine present in the cloud-top levels, but it is believed that this originates in the deep atmosphere of the planet. When ultraviolet sunlight is incident upon the phosphine, red phosphorus is produced. So the red spot would appear to be another region of strong upward motion. The presence of such a feature in Saturn's atmosphere provides further evidence that red spots are naturally occurring features in the gaseous atmospheres of the giant planets.

The Voyager observations did not permit a detailed study of the southern hemisphere at the resolution that was possible for the northern hemisphere. However, a ribbon-like feature was observed similar to that in the northern hemisphere, which suggests that there may be some symmetry in the cloud morphology as the zonal wind profile indicates.

Long-lived features

The atmospheres of Jupiter and Saturn both possess examples of cloud systems that seem to persist for months, years, even decades and centuries. Eddies in the oceans and atmosphere of Earth are much less enduring by several orders of magnitude. In the Atlantic Ocean eddies tend to drift in an easterly direction until they merge with the Gulf Stream off the east coast of North America, and their lifetimes are measured in months and sometimes even years. In the atmosphere, the long-lived eddies are generally associated with specific surface features, such as mountain chains, or boundaries between continent and ocean. On Saturn and Jupiter, however, there is no such topography to generate eddies and flow patterns similar to those on Earth.

Two possible explanations have been suggested to account for the long-lived behavior and interactions of cloud features in their planetary atmospheres. One suggestion is that we are observing a "solitary wave", which is a self-containing wave with a single crest instead of a train of crests and troughs. Such a system could produce a flow pattern similar to that observed around Jupiter's Great Red Spot. However, when two solitary waves meet, they simply pass through each other, while observations show that on Jupiter and Saturn the ovals sometimes merge.

An alternative proposal assumes that the east–west flow pattern of the clouds is related to a much deeper system, perhaps a system of rotating cylinders (*see* page 25). It is also assumed that the vortex extends downwards only to the top of the adiabatic zone, and that the stability of the vertical extent of the flow enables the vortices to survive even large-scale perturbations. Large spots may grow by consuming smaller ones, which are transient in behavior.

Weather and color

An overall understanding of the meteorological processes on Saturn and Jupiter will depend upon determining the depth to which the motions extend, and the individual roles of internal and solar heating in driving those flows.

One of the fundamental problems associated with Jupiter and Saturn is to explain the origin of the colors in their atmosphere, and the apparent differences between the two. The Earth's clouds are composed of water, and particles of water ice that are white. The colorful appearance of the clouds on the two giant planets provides vital evidence of the differences in the chemical composition of their atmospheres. Color is caused by a disturbance to the chemical equilibrium, by charged particles or vertical motion for example.

1A

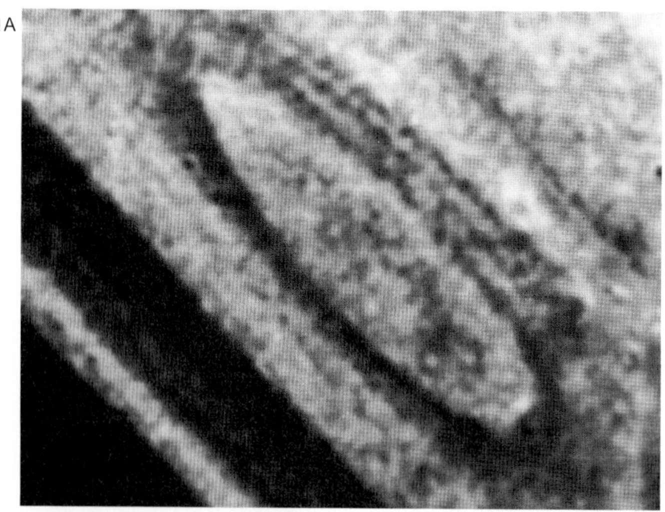

B

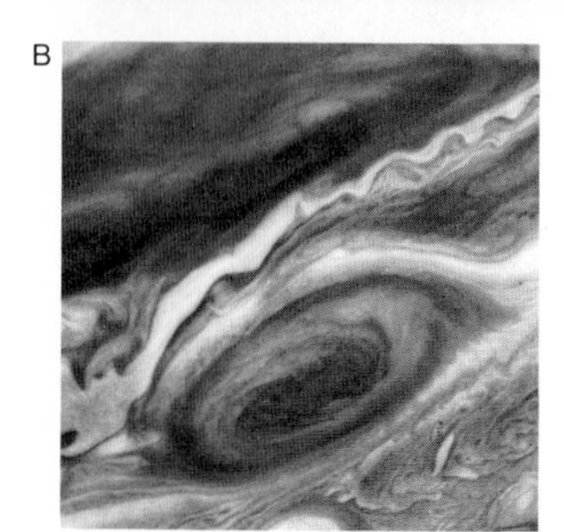

1. Anne's Spot
In Saturn's southern hemisphere at 55° latitude, Anne's Spot (**A**) is red in color, rather like Jupiter's Great Red Spot. Saturn's feature is about 3,000 km in diameter and it moves in an easterly direction at 30 ms^{-1}. This Voyager 2 image was taken through a green filter two days before encounter. The Great Red Spot on Jupiter (**B**) is a much larger feature. We know it is much colder than its surroundings and its color may come from phosphorus.

2

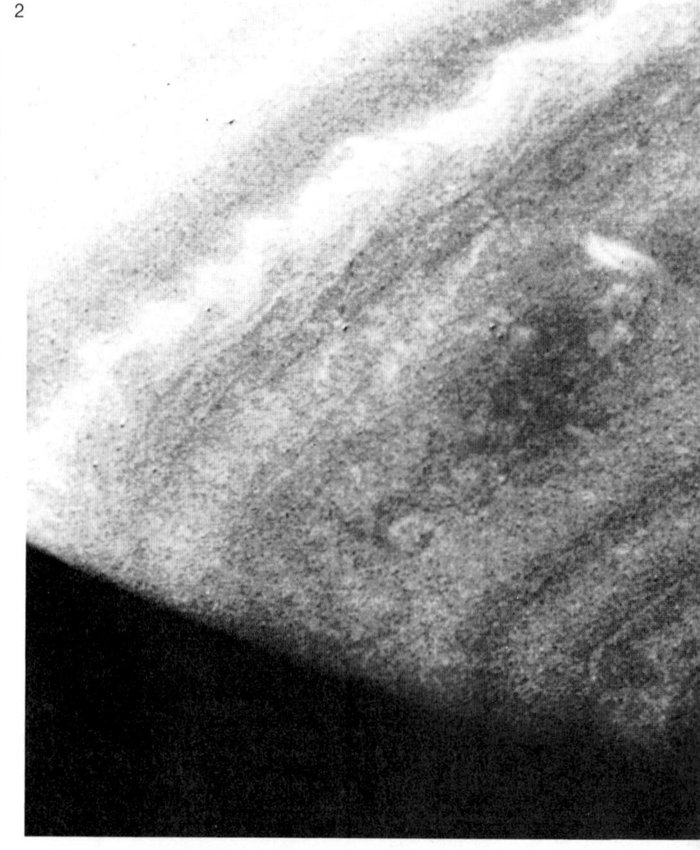

Altitude reflects the processes that cause chemical reactions in the first place: there are more charged particles at higher levels of the atmosphere as well as more sunlight. Ammonia, which is known to be a constituent of the Saturnian (and Jovian) atmosphere, condenses into a layer of cloud at a particular level in the troposphere; and it is believed that ammonia clouds constitute the regions of white clouds in the Saturnian and Jovian atmospheres. Saturn's atmosphere is colder than Jupiter's, and there are more extensive clouds of ammonia. Beneath the clouds of ammonia other layers are believed to form, possibly including ammonia hydrosulphide (NH_4SH), water ice and ammonia solution.

Saturn's pastel colors

As temperatures decrease with increasing height, different colors are seen. On Saturn, however, the colors seem more pastel than those observed on Jupiter. It is possible that diffuse layers of photochemically produced aerosol products are distributed throughout the upper atmosphere of Saturn, forming a thick haze that obscures both features and colors underneath. Consequently, this extra layer of material would reduce the contrast in the colors. Furthermore, it is possible that polymer substances produced in the clouds generate varied colors at these lower temperatures. There may just be a lack of atmospheric colored compounds (chromophores). Chromophores may be generated at a lower rate than the rate of the zonal wind flow so that eddies, for instance, are dispersed more quickly than chromophores can form, which could account for Saturn's bland appearance. The planet's lower temperature than Jupiter's, particularly at cloud-top levels, could mean that chemical reactions are slower, the result of which could be a more chemically homogeneous atmosphere with a more uniform distribution of chromophores. All these factors could contribute to the pastel appearance of Saturn.

Saturn's atmosphere is known to include small but important amounts of hydrocarbons, such as methane, acetylene and other exotic products. The clouds are bathed in ultraviolet sunlight, which is a strong source of energy for chemical reactions. There are also likely to be violent lightning flashes, which, like terrestrial storms, are associated with regions of strong convective activity. These storms extend throughout the clouds, and they provide a second important source of energy to affect the complicated atmospheric chemistry at a variety of levels.

Origin of the colors

The colorful clouds, which tend to lie in broad longitudinal bands, are thought to be created from a mixture of methane, ammonia and probably sulphur too. Molecules are broken up by ultraviolet sunlight and violent lightning, and then stirred up by the dynamic weather patterns. A possible key to understanding the diversity of colors would be the identification of sulphur-bearing substances, since sulphur and certain of its compounds produce a range of colors from yellow to brown or black depending on the temperature. A suitable candidate is hydrogen sulphide (H_2S), from which hydrogen polysulphide (H_xS_y) or ammonium polysulphide ($(NH_4)_xS_y$) might form. These chemicals are capable of producing the colors that are to be seen on Saturn, but other compounds—yet to be detected in the planet's atmosphere—may be responsible for some of the colors.

The color of the red spot on Saturn may be related to the presence of phosphine in the atmosphere; for this to be possible, red spots must penetrate more deeply into the atmosphere than other features and they probably represent regions of strong upward motion. It is possible that hydrocarbons such as acetylene and ethane act as scavengers to the phosphine reactions, reducing the amount produced. Variations in the concentrations of these hydrocarbons and therefore of phosphine and other products may help to explain the variety of colors and shades.

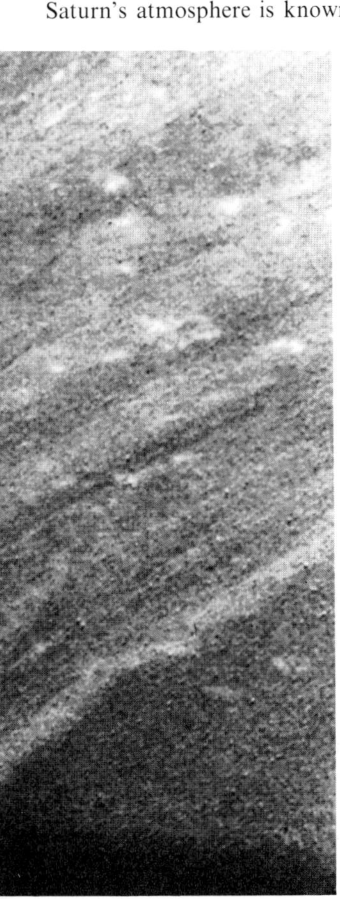

2. Southern hemisphere
From a distance of 442,000 km Voyager 1 took this wide-angle photograph of the south polar region and the mid-southern latitudes of Saturn. Again the light and dark bands can be seen to be made up of many small-scale features such as waves and eddies. Saturn's equivalent of Jupiter's Red Spot can also be seen. The Voyagers' scrutiny of the southern hemisphere was less detailed than their survey of the northern hemisphere.

3. Cloud layers
All the observable features on the Voyager images correspond to clouds of various colors and brightness. Infrared observations show that the bands of color on Saturn distinguish different levels in the planet's cloud structure. Blue clouds have the highest brightness temperature so they must lie at the deepest levels in the atmosphere. We see them only through gaps in higher cloud which permit us to look down. Brown clouds are the next highest, and finally white clouds. If there are any red clouds, as in Jupiter's Great Red Spot, they are at the highest level and represent very cold features as indicated by infrared readings. In the case of Saturn, there is probably a haze layer above the clouds and that this results in muted colors.

3

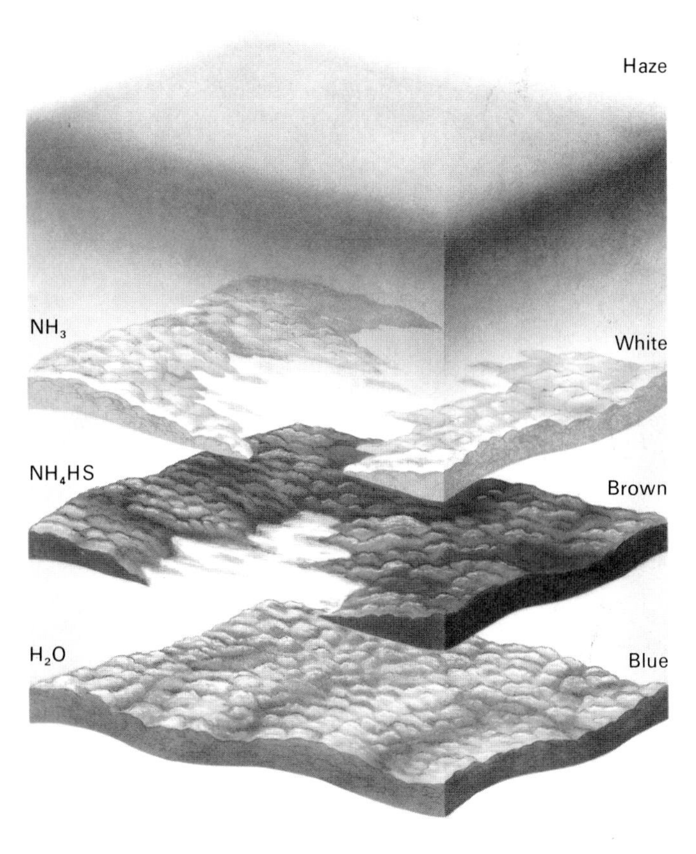

The Rings Introduction

Saturn's system of rings is no longer unique, in as much as both Jupiter and Uranus have rings associated with them; but those of Saturn are in a class of their own. In particular they are bright, with a maximum albedo (*see* Glossary) higher than that of Saturn's disc, which is why the planet appears much brighter, as seen from Earth, when the rings are wide open than when they are edgewise-on. Edgewise presentations occur at alternate intervals of 13.75 yr and 15.75 yr. The inequality is due to Saturn's eccentric orbit.

The rings lie in the plane of Saturn's equator, which is tilted at an angle of 28°. During an edgewise presentation the rings are hard to see even with powerful telescopes, because they are thin and because they are barely illuminated by the Sun. There are times when the Sun and the Earth lie to opposite sides of the ring-plane, so that the unilluminated ring-face is presented. The rings lie inside the Roche limit for Saturn; that is, the minimum distance from the planet at which a body with virtually no gravitational cohesion could survive without being disrupted.

General characteristics

There are three main rings: two of these are bright, and are lettered A (the outer) and B; they are separated by the Cassini Division, named in honor of G. D. Cassini, who discovered it. Closer in to the planet is Ring C, also known as the Crêpe or Dusky Ring, which is semitransparent and is less evident. The main rings together measure about 275,000 km in width.

New rings have been found from instruments carried on the space-probes. Pioneer 11 detected Ring F, 3,600 km beyond the outer edge of Ring A, and the Voyagers added two more; Ring G, which is excessively tenuous and has been seen only in forward-scattering light, and the very broad and also very tenuous Ring E, which extends out into the main satellite system.

Pioneer 11 provided a tantalizing first glimpse of the ring system at close quarters in transmitted rather than the more usual reflected sunlight, but it was the Voyagers that showed many unexpected features. It had been thought that the rings would be of fairly straightforward structure: several well-defined rings separated by gaps which could be attributed to the gravitational effects of inner satellites such as Mimas. The Voyagers showed that there were literally thousands of rings, with minor divisions.

Information about the composition of the ring particles can be derived from the rings' ability to reflect or absorb light at different wavelengths. The A, B and C rings are poor reflectors of sunlight at certain near-infrared wavelengths, which indicates water ice.

The sizes of the particles range from tiny "grains" to blocks several tens of meters in diameter. Observations at radar wavelengths provide information on the size of the particles. The high reflectivity of the A and B rings implies that most particles are at least comparable in size to the radar wavelengths of several centimeters. Radar observations have set the upper size limit of particles, while observations of the scattering of sunlight at visible wavelengths show smaller particles—of the order of a micrometer. The particles are less reflective in blue than in red light, perhaps because of other substances present. Dust containing iron oxide may be one possible source of the reddish color. The gaps are not true gaps; even the Cassini Division was found to contain numerous ringlets.

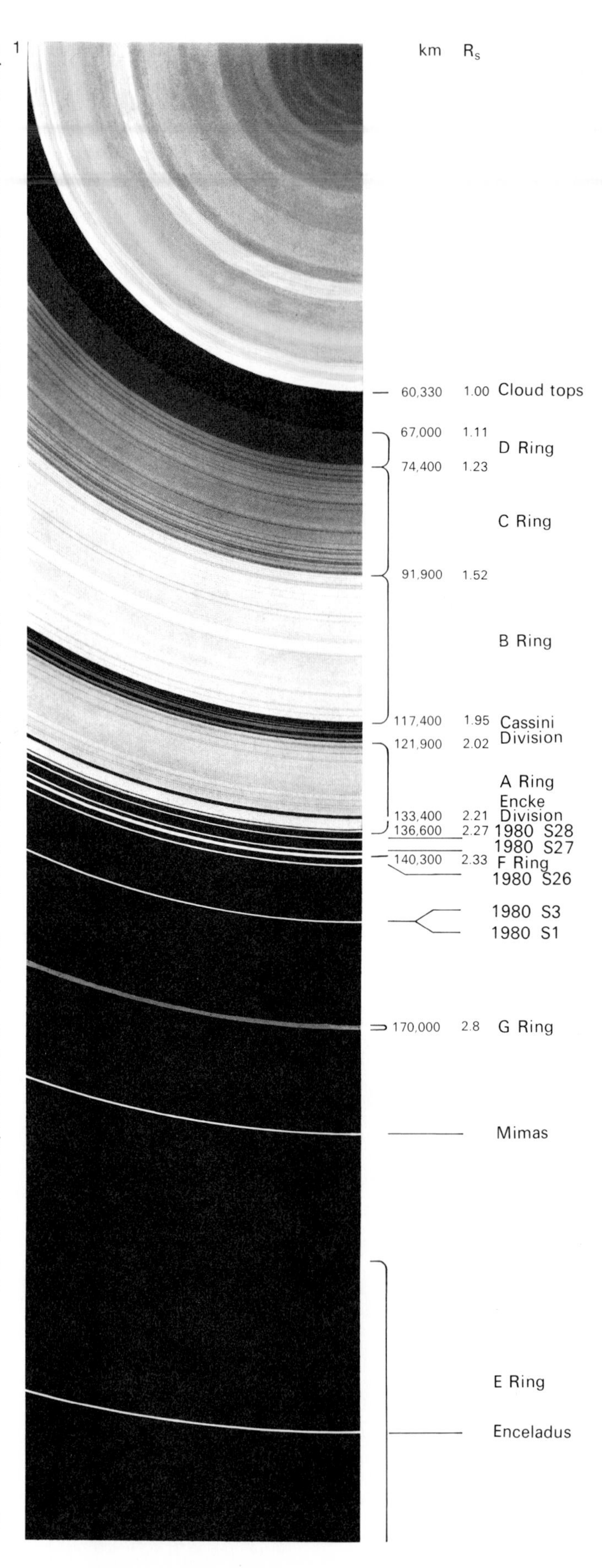

1. The ring system
The D Ring, closest to Saturn, extends from 12,700 km (1.2R_s) above the planet to the cloud tops or the upper atmosphere. The C Ring is in fact dozens of ringlets, at least one of which is eccentric. It is comparatively transparent, and extends to 1.53R_s. The next ring out—the B Ring—has the spokes, and extends outward as far as 1.95R_s. It is separated from the A Ring by the Cassini Division. The A Ring extends from 2.01R_s to 2.26R_s and contains the Encke Division. Beyond the classical rings lie the F Ring (2.33R_s), G Ring (2.8R_s) and the E Ring at 3.5R_s extending as far as 3.9R_s.

2. D Ring
This Voyager 2 image of the D Ring was recorded at a range of 250,000 km. It is an extremely faint ring and is composed of several bands where ring particles are concentrated. The limb of Saturn is seen in the upper left, and the planet's shadow cuts diagonally across the image. The D Ring's outer edge is the C Ring's inner border.

3. Reflectivity
Comparative shots of the dark and illuminated sides of Saturn's rings show a reversing out of the system's components. On the illuminated side, material in the C Ring and the Cassini Division is not apparent, while the A and B rings are bright in forward-scattering light, indicating a significant proportion of very small particles. On the dark side the optically thin C Ring and Cassini are bright as a result of diffuse scattering.

4. Detail of the B Ring
This Voyager 2 detail of the B Ring was taken at a distance of less than 800,000 km and it revealed about ten times more ringlets than had previously been suspected in the B Ring. The finest rings to be seen on this image are about 15 km across.

Ring D
It is hardly correct to describe this as a true ring, and it is better termed the D region. It is very thin, though there are various narrow components. There may be no well-defined inner edge, in which case the D region seems to extend right down to the cloud tops of Saturn, though it is also possible that there is a lower limit at a distance of about 6,450 km from the uppermost clouds. The D Ring has been seen only in forward-scattering light, and, due to the intrinsic brightness of Saturn, it could not be seen from Earth.

Ring C
The Crêpe Ring fills the area between the D region and the inner edge of the brightest ring, B, at 25,000 km above the cloud tops. It is a very complicated, grooved region, with a large number of discrete ringlets which, as seen from Earth, merge together. This was shown by Voyager 1, and even more dramatically by Voyager 2, which provided higher-resolution images. The structure of the C Ring is regularly ordered and cannot be readily associated with gravitational resonances. There are at least two particularly noticeable gaps, the outer of which is about 270 km wide, and is flanked symmetrically by several sharply bordered bands, which are less transparent than their surroundings. This gap is not empty. It contains a narrow ringlet, which is eccentric in shape rather than being perfectly circular. It is about 35 km wide and 100 km from the outer edge of the gap at its periapse (nearest to the planet) and 90 km wide and 50 km from the outer edge of the gap at apapse (furthest from the planet). Eccentric rings came as a great surprise; there are quite a number of them, and they cannot yet be fully explained. The eccentricity may be caused by a small satellite or satellites in the vicinity; by gravitational instabilities within the rings; or by density waves caused by satellite resonances.

Ring B
Ring B is the brightest part of the entire ring system, and is opaque when seen from Earth. It had been expected to be simple in structure, but was instead highly complex. Voyager 1 found hundreds of ringlets and minor divisions, while Voyager 2 increased this number to thousands. The inner part of the B Ring contains the most numerous and transparent gaps. Near the innermost edge the ring tends to be more opaque and to resemble the extremely opaque outer third quarter of the ring. These similar regions have fewer narrow gaps. The most opaque regions have the spokes.

The thickness of the ring system is presumably at its maximum in Ring B, and before the Voyager pass estimates had ranged from a few millimeters to as much as 15–20 km. It now appears that the true value is between 100 and 150 meters, which explains why the rings almost disappear when seen edge-on. The Voyagers also showed that there is a large, rarefied cloud of neutral hydrogen extending to about 60,000 km above and below the ring-plane and beyond the outer edge of Ring A. Water ice in the ring particles is presumably the source; it has been estimated that the density of the hydrogen is 600 atoms per cubic centimeter.

Images of the B Ring are particularly spectacular, though it must be borne in mind that the actual images from Voyager are taken in black and white; the colors are produced by the use of suitable filters. When the images are computerized it is possible to reproduce faithfully the actual colors, though in many cases the colors are either enhanced or modified to make analysis easier.

The particles making up the B Ring are not identical in composition with those of the C Ring or the D region. In particular they are distinctly redder, and the average particle size is probably of the order of several centimeters to a few meters in diameter. The B Ring extends altogether from 25,000 to 54,000 km above the uppermost clouds. It thus extends from the outer boundary of the C Ring to the inner edge of the Cassini Division. The extremely sharp boundary between the B and C rings does not show any gaps.

The Rings Spokes

General characteristics

The radial features, or spokes, detected by Voyager 1 and confirmed by Voyager 2 in the B Ring have proved to be among the most interesting features of Saturn's rings. They were completely unexpected and have not yet been adequately explained. They appear dark in back-scattering light and bright in forward-scattering illumination. Observations indicate that small particles constitute a large proportion of the B Ring and are more visible perhaps because they are elevated above the ring plane within the spokes. Most of the spokes are confined to the central B Ring in a region ranging from 43,000 to 57,000 km above Saturn's cloud tops. There is no sign of them closer to Saturn, across the Crêpe Ring, or in Ring A. Many of them are 10,000 km long by 2,000 km wide. At all times during the Voyager 1 and 2 encounters there were spokes on view in Ring B. In each case the results were combined to make "moving pictures" of the spokes which were remarkably interesting to those trying to explain the phenomenon.

The spokes are essentially radial. They are thought to form radially, in a frame of reference rotating closely to the co-rotational rate of Saturn's magnetic field, and then to follow the differential orbital motion of the individual ring particles. The rotation period of the inner edge of Ring B is 7.93 hr and that of the outer edge 11.41 hr. So, in accordance with Kepler's laws, particles in the inner regions of the ring have shorter periods than those in the outer regions. Consequently, no radial features should exist. There is no apparent reason why they should form, and if they do, they should dissipate after a very short period of time. Yet the spokes in Ring B persist for hours. They are distorted by the rotation, but they do not disappear as quickly as might be expected.

Formation of spokes

It appears that the magnetic field is responsible for the formation of the spokes and that Keplerian motion is responsible for the dynamics of the particles. Some of the spokes appear to be wedge shaped—with their broadest end towards Saturn—and this shape may reflect the difference between Keplerian and magnetic orbital motion over the time it takes for one spoke feature to be formed. The time scale for the formation of a 6,000 km feature may be as short as five minutes. "Young" spokes are close to the co-rotational rate of the magnetosphere, whereas "older" spokes have a Keplerian rate. These are inevitably distorted and are termed "non-radial" by some astronomers, although this term could be misleading and is not an official categorization.

Some spokes have been tracked as they rotate through 360° or more. It is not clear, however, whether the same spoke pattern is observed throughout that period, or whether a new one is reprinted on top of the old one. It is not thought that the lifetime of any one spoke is very long.

Dark spokes

Spokes are also seen on the unilluminated side of the B Ring, which is illuminated predominantly by sunlight scattered off Saturn's atmosphere. It is believed that these are dark-side phenomena and not bright spokes shining through optically thin parts of the B Ring. The shapes seen on the dark side suggest that the spokes follow similar morphological behavior as is observed on the illuminated side. The fact that these features must be created on the dark side of the B Ring while others, it is thought, form in Saturn's shadow has implications for the mechanism of spoke formation and casts doubt on the theory that photoionization of small particles levitates them out of the ring-plane. The detection of electrostatic charges (lightning) at radio wavelengths from the ring does suggest that this mechanism is related to the spokes. Voyager 2's detection of discharges was only 10 percent that of Voyager 1, however, indicating that conditions within the ring system are constantly changing.

1

2

1. Rotation of spokes
Time-lapse photography showed not only a relatively short time in which spokes are formed, but also the co-rotation of spokes with the rings. The spokes' ability to do this is believed to be associated with Saturn's magnetic field.

2. Bright-side spokes
One of a sequence of Voyager 2 shots of the B Ring shows several spokes with a variety of widths and tilts. The Cassini Division is visible at the left-hand side of the photograph.

3. Dark-side spokes
Spokes seen from the underside of Saturn's rings appear bright because they are illuminated by forward-scattering light reflected off the planet's disc.

3

Color Plates

Saturn, Tethys and Dione
Voyager 1
3 November 1980

Saturn
Catalina Observatory (top)
11 March 1974

Saturn
Pioneer 11 (bottom)
29 August 1979

Saturn
Voyager 1 (top)
30 October 1980

Saturn
Voyager 2 (bottom)
4 August 1981

Saturn
Voyager 1
16 November 1980

Northern Hemisphere
Voyager 2
19 August 1981

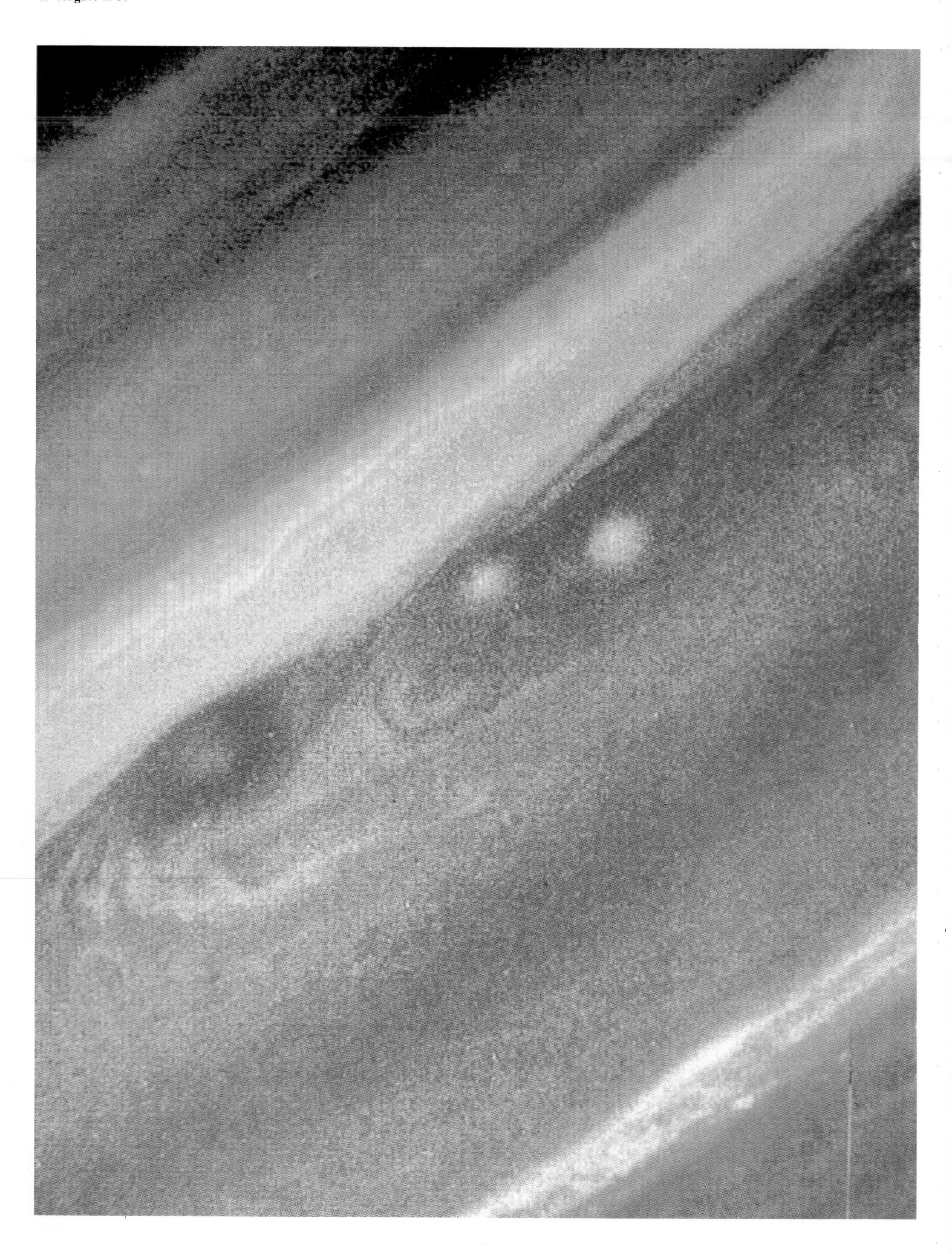

North Polar Region (top left)
Voyager 2
25 August 1981

Northern Hemisphere (top right)
Voyager 2
19 August 1981

Northern Hemisphere (bottom left)
Voyager 1
5 November 1980

Red Oval
Voyager 1
6 November 1980

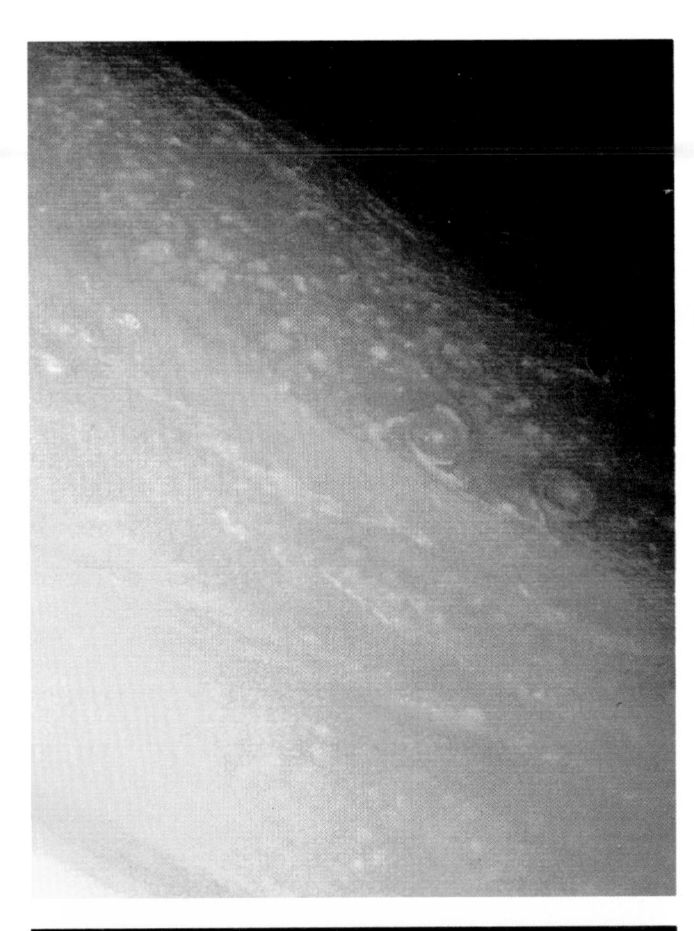

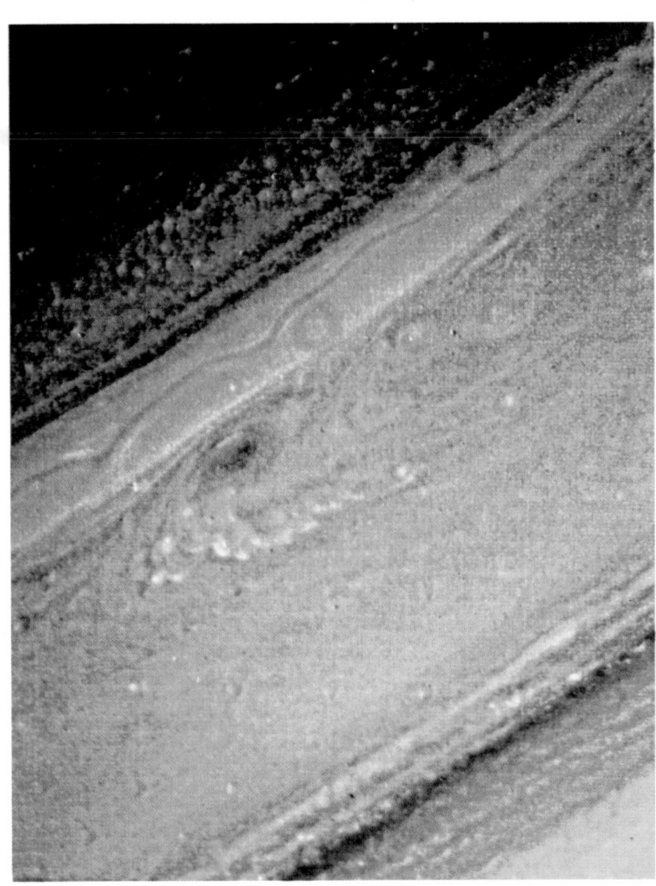

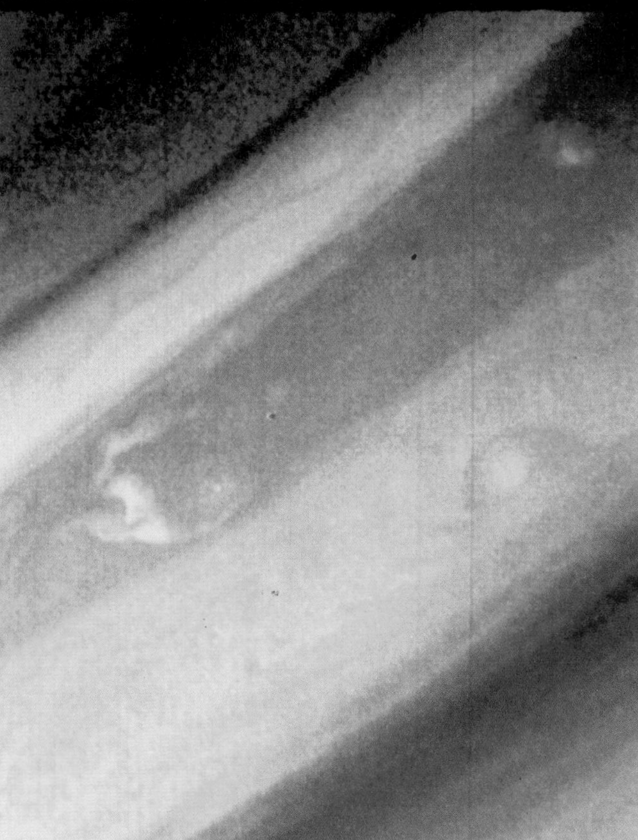

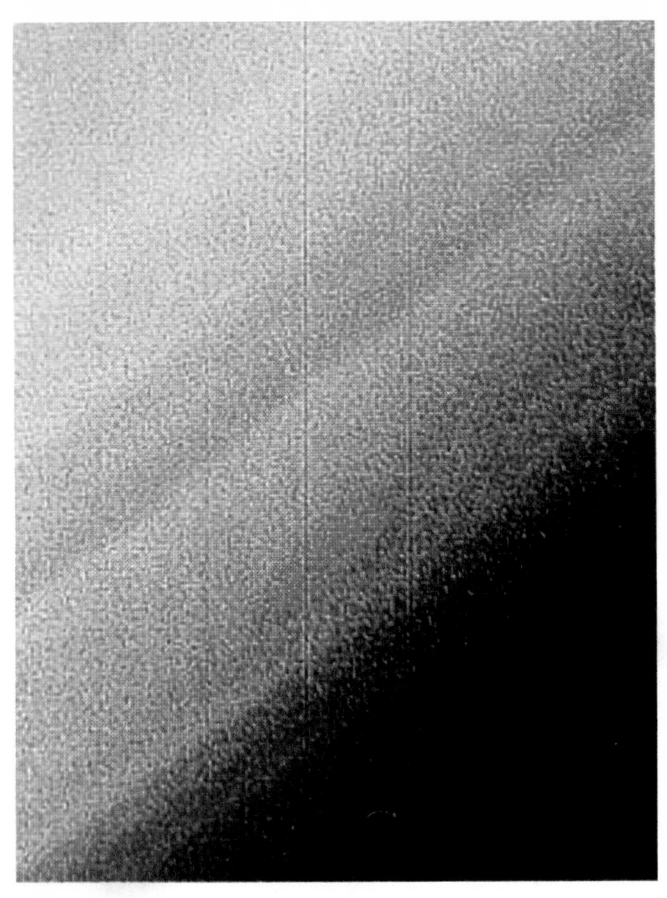

The Rings
Voyager 2 (top)
23 August 1981

The Rings
Voyager 1 (bottom)
12 November 1980

Titan
Voyager 1 (top left)
9 November 1980

Voyager 2 (top right)
25 August 1981

Titan limb (bottom)
Voyager 1
12 November 1980

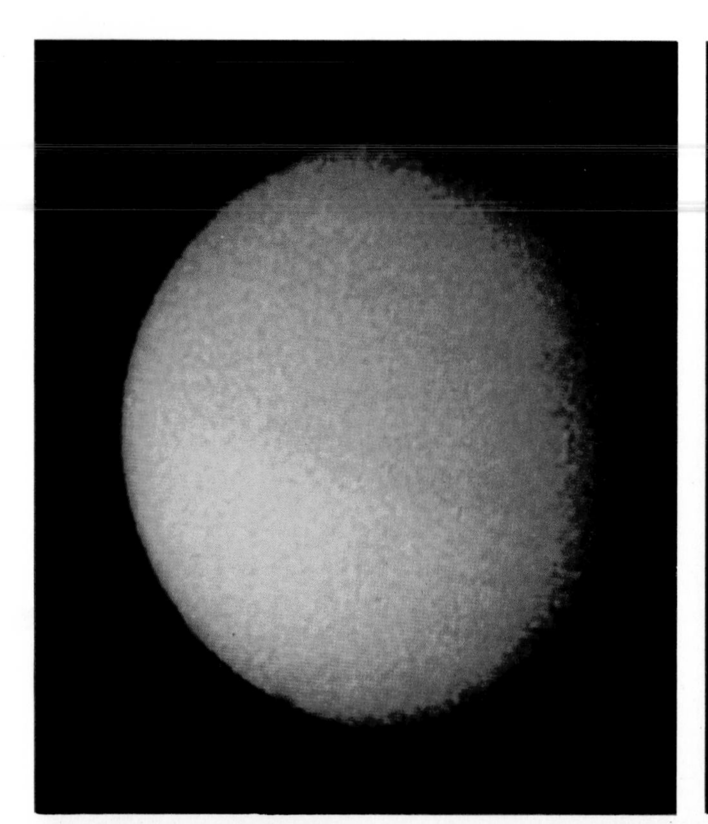

Mimas
Voyager 1
12 November 1980

Enceladus
Voyager 2
25 August 1981

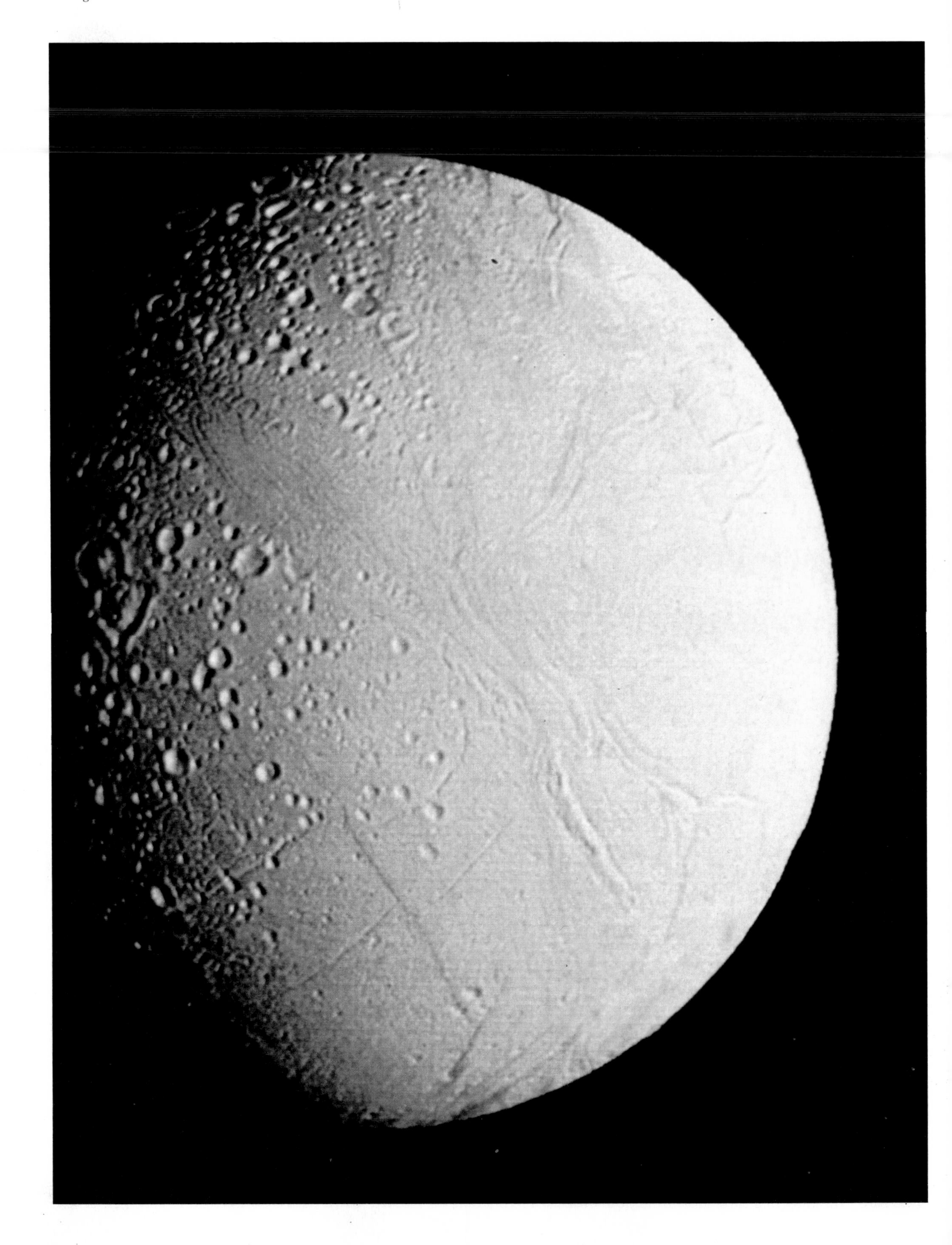

Tethys
Voyager 2
25 August 1981

Dione
Voyager 1
12 November 1980

Rhea
Voyager 1
12 November 1980

Iapetus
Voyager 2
22 August 1981

Notes to Color Plates

Page 33. Saturn and two of its moons, Tethys and Dione, from a distance of 13 million km. One of the moons has cast a shadow on the cloud tops just below the rings in the image.

Page 34. View of Saturn from Earth in 1974 (top), with the 1.5 m telescope at Catalina Observatory in Arizona. A 1979 Pioneer 11 photograph (bottom) shows light scattered through the ring system.

Page 35. The Voyager 1 image (top) is contrast-enhanced and shows Saturn in back-scattered light. The Voyager 2 image, from a distance of 21 million km, is similarly enhanced (bottom).

Pages 36–37. Four days after its encounter with Saturn, Voyager 1 looked back on the planet from a distance of more than 5 million km and saw Saturn as a crescent.

Page 38. Three spots in Saturn's northern hemisphere are visible in this false-color image from Voyager 2. The largest one is 3,000 km in diameter. To the north of them is a ribbon-like feature.

Page 39. Photographs of Saturn's North Polar Region (top left) and Northern Hemisphere (top right) show a variety of features, many of which indicate motion within the clouds. The North Temperate Belt (bottom left) also appears to be active. This image is color-enhanced and shows convective features. Anne's Spot in the southern hemisphere (below right) is a miniature version of Jupiter's Great Red Spot.

Pages 40–41. A series of false-color images from Voyager 2 (top) from a range of 2.8 million km were processed to enhance color differences between the rings and give clues about compositional variations. Voyager 1 took dramatic shots of the underside of the rings from a range of more than 700,000 km (bottom).

Page 42. Titan from Voyager 1 (top left), photographed three days before the spacecraft passed within 7,000 km of the satellite, showed a slightly brighter southern hemisphere. When Voyager 2 photographed Titan from 2.3 million km there was a dark north polar ring. False-color processing of Titan's limb (bottom) reveals the atmosphere as a blue haze.

Page 43. The south polar region of Mimas shows a very heavily cratered terrain.

Page 44. A dramatic variety of terrains can be seen on Enceladus, which, viewed from 119,000 km, resembles Ganymede in Jupiter's satellite system.

Page 45. A long trench on Tethys winds its way through cratered regions on this moon that is practically all water ice.

Page 46. Dione is another moon that is intensely cratered but also exhibits sinuous valleys and bright wispy streaks.

Page 47. Voyager 1 was only 70,000 km above the north pole of Rhea when this photograph was taken. Features as small as 2 km across can be seen.

Page 48. The surface of Iapetus shows both very bright and very dark areas, the nature of which is still subject to debate.

The Rings Structure I

The Cassini Division

The Cassini Division is one of the most prominent features of Saturn's ring system. When the rings are wide open, as they will be in the mid-1980s, a very small telescope will show the Division, and even when the rings are tilted at a less favorable angle the dark line can generally be seen near the ansae, or ring-tips. Before the space-probe encounters, the Cassini Division was assumed to be a true gap, virtually devoid of ring particles. At one time there was even talk of sending Pioneer 11 through it.

The Division is of considerable width, nearly 4,000 km—as wide as the North American continent. Its material is similar to that of the Crêpe Ring; the particles are decidedly less red than those of Ring B and there is a dearth of small particles. Both the Cassini Division and the C Ring exhibit discrete, regularly spaced bands of uniform brightness, together with a variety of narrow, sharp-edged, empty gaps with a radial width of 50–350 km. Some of the gaps contain even narrower, equally sharp-edged ringlets that are quite opaque. These differ in color from the surrounding ring material and resemble more closely the material in the A and B rings.

The origin of the Division has to be considered together with the lesser divisions in the ring system. It is worth noting that there is something strange about the outer edge of Ring B, which is also the boundary of the Cassini Division. Its distance from Saturn is not constant, but varies perceptibly to the extent of about 140 km. Moreover, although it is elliptical, it does not obey Kepler's Laws, because Saturn lies at the center of the ellipse, not at one of the foci.

According to Kepler's Laws, a planet moving around the Sun will travel in an elliptical orbit; the Sun occupies one of the foci of the ellipse, while the other focus is empty. The same applies to a satellite in orbit around a planet, though the situation is complicated by the presence of several large satellites, as with Jupiter in particular and Saturn to a lesser degree. The particles in the B Ring should behave essentially like tiny individual satellites do; each moves around the planet in its own independent path, and should therefore follow a Keplerian orbit. This, however, is not the case. It may be significant that the particles at the outer edge of Ring B have a period which is half that of Mimas, the innermost of the main satellites; the periods are 11.41 hr and 23.14 hr respectively.

Ring A

The A Ring also contains an important division. The A Ring is not as bright as its inner neighbor, and the difference is quite marked, even from Earth. Ring A is also much less opaque, though the inner edge is as sharply defined as the boundary between Ring B and the Crêpe Ring. In Ring A, as with Ring B, there are large numbers of ringlets and minor gaps, quite apart from the principal division, popularly known as the Encke Division, but now officially called the A Ring Gap.

As Voyager 2 drew away from Saturn there was a period when a bright star, Delta Scorpii, lay on the opposite side of the ring-plane, and was therefore occulted by the rings as seen from the spacecraft. Continuous measurements were made, Delta Scorpii being hidden every time it passed behind a ringlet and reappearing every time its light shone through a gap. From the results it would seem that there are very few clear gaps anywhere in the ring system. The Encke Division, like the Cassini Division, contains several ringlets, at least two of which are eccentric. There are two apparently discontinuous rings inside the Division. The clumpiness of these rings can be seen to orbit as a pattern at the orbital period of the rings at that radius. Both rings vary in brightness and one has a kinked morphology, maybe caused by perturbations created by eccentric satellites. The width of the Division is only 200 km, but even so it is visible with a moderate Earth-based telescope under good conditions. It lies about 3,000 km inside the outer edge of the A Ring. Outward of the A Ring Gap there is a pattern of unresolved ring features. They may represent a sequence of classical resonances converging on one of

1

1. Rings of gold
This Voyager 1 picture was taken shortly after the spacecraft's closest approach to Saturn in November 1980. It clearly shows the Cassini and Encke divisions as well as, at middle left, the opaque nature of the B Ring.

2. Cassini Division
The Cassini Division has been shown to have a very complex structure. From the outer edge inward in this image can be seen a medium-dark ringlet 800 km wide; four brighter ringlets about 500 km wide and separated by dark divisions; and a hardly visible, narrow but bright ringlet at the inner edge, and little more than 100 km wide. The entire Cassini Division is defined as that area between the two dark ringlets and was photographed by Voyager 1 at a distance of 6 million km from the ring system.

3. B Ring variations
Radial slices through the Cassini Division (top) and the B Ring (bottom) show variations of up to 140 km in the width of the gap separating the two, believed to be the result of resonances between the ring material and the satellite Mimas. The presence of an eccentric ringlet can also be detected within the gap.

2

3

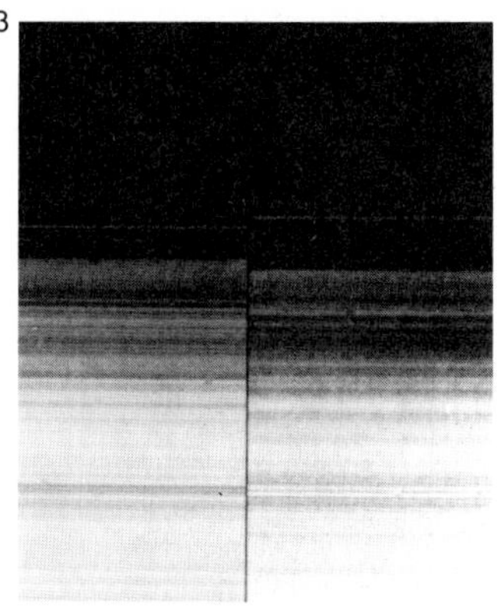

4. Resonances
Resonant orbits are a feature of several regions of the Solar System. A resonance occurs between two objects where the orbital period of one is commensurate with the orbital period of the other. A resonant position may be vacant or packed with objects. Kirkwood Gaps in the asteroid belt occur where asteroids would have orbital periods that are simple fractions of the period of Jupiter, which dominates the asteroid zone. The Cassini Division is like a Kirkwood Gap; a particle in the Division has a period about one half of the satellite Mimas. Several gaps in Saturn's A Ring have resonant locations with S10 and S11, and the edges of rings A and B are near resonant locations.

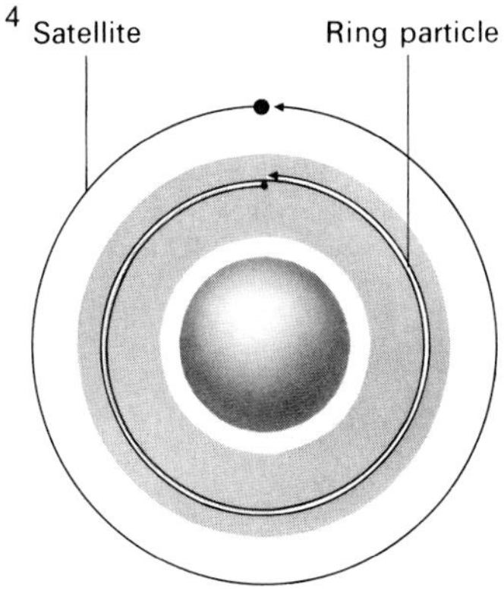

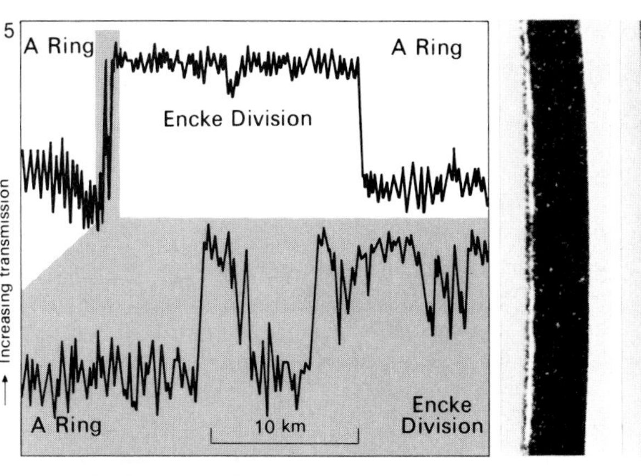

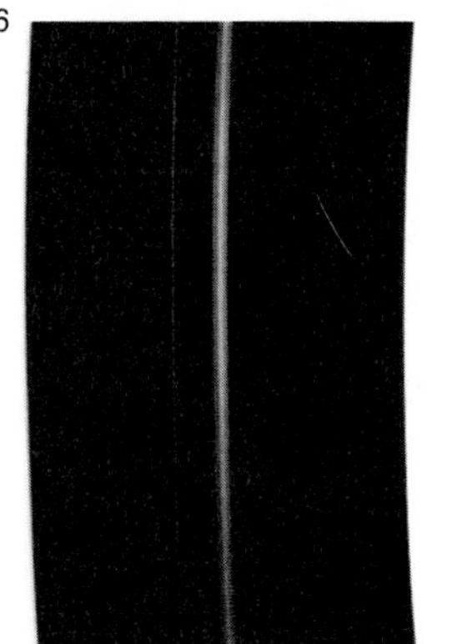

5. Structure of the Encke Division
The graphs show the amount of light recorded from Delta Scorpii as the ring system passed through the line of sight. The Division's structure was revealed, including a feature near the inner edge. The upper graph was averaged to a 3.5 km resolution, the lower 300 m.

6. The Encke Doodle
The irregular ringlet in the Encke Division was nicknamed the Encke Doodle. First seen by Voyager 1, it was observed more closely by Voyager 2. There are believed now to be two ringlets: at the center and the inner edge.

the inner satellites. The entire region outside the A Ring Gap is about 25 percent brighter than the region inside. A similar situation is found in a region at the outer edge of the A Ring, several hundreds of kilometers wide, that is set off from the A Ring by an apparently vacant gap tens of kilometers wide. It appears to be about 50 percent brighter in forward-scattering light than the region inside the gap. The outer edge of the A Ring is very sharp, which may be related to the proximity of the satellite 1980 S28.

Causes of gaps

Before the Voyager results, it was regarded as virtually certain that the gaps in the ring system were due to the gravitational perturbations of the satellites, particularly Mimas and Enceladus, which were the closest members of the satellite family then known. It was assumed that if a ring particle has a revolution period that is an exact fraction of that of a satellite, it would be subject to cumulative perturbations, and would be moved out of the relevant area in a fairly short time. When ring particles are resonating with a nearby satellite, they begin to swing back and forth, nearer and further from Saturn. The effect is rather like pushing a swing: if the swing is large enough the swinging will increase. Mimas, for example, exerts a gravitational push on the particles in the ring system. This swinging causes particles to collide and certain areas are therefore "cleared out". In other words, the satellites would keep the Cassini and Encke divisions "swept clean", and no ring particle entering these divisions could stay there for long.

There were analogies to be found in the asteroid belt. This lies between the orbits of Mars and Jupiter, and contains many thousands of small worlds, most of them only a few kilometers or tens of kilometers in diameter. There are certain well-defined gaps where few or no asteroids are to be found; they are known as the Kirkwood Gaps. An asteroid moving in a Kirkwood Gap would have a revolution period that was an exact fraction of that of Jupiter, and so it would be unable to stay in such an orbit.

This was all very well as far as Saturn's rings were concerned when it appeared that there were only a few divisions. There are only a few strong resonances, however, and not enough to explain the highly complicated structure of the ring system, which involves thousands of what may be regarded as divisions even though they are by no means devoid of ring particles. There is evidence that satellite perturbations play a role, but there must be other influences as well.

The current theory concerns what may be termed density waves, which are formed by the gravitational effects of the satellites. The Voyager observations show a sequence of smoothly undulating brightness fluctuations in the outer Cassini Division within a band located between about 120,700 and 121,900 km from the center of Saturn. The observed behavior of these variations suggests that they are spiral density waves caused by resonances between local ring particles and Saturn's outermost large satellite, Iapetus. These spread outward from positions where the revolution periods of ring particles are exact fractions of that of one of the satellites, rather as a moving ship produces a wake in a calm sea.

It appears that at the distance at which a ring particle would have a period of one half that of the satellite 1980 S1 there is a series of outward-propagating density waves, with characteristics indicating that there are about 60 grams of ring material per square centimeter of ring area and that the relative velocities of the particles to each other is of the order of one millimeter per second. Other examples could be cited, but this explanation is probably the best available, even though it is far from complete.

Finally, it is too early to say whether the minor ringlets and divisions shown by the Voyagers are permanent or not. They may be no more than transitory, though it seems likely that the Cassini Division is a genuinely permanent feature of the ring system, and the same may well be true of the Encke Division in Ring A.

The Rings Structure II

The main ring system ends at the outer boundary of Ring A, at a distance of 73,000 km from Saturn's cloud tops or about 136,000 km from the center of the planet. There are three more rings further out, lettered F, G and E. They are faint, and without space-probes it is not likely that they would have been detected.

Ring F

Ring F was detected by particle absorption experiments on board Pioneer 11 in 1979, and caused tremendous interest when Voyager 1 approached Saturn and sent back pictures showing that Ring F was compound. It appeared to be made up of three separate strands, which were intertwined; it was nicknamed the "braided" ring, and seemed to defy all the known laws of dynamics. When Voyager 2 made its pass the aspect of Ring F had changed. There were five separate strands in a region that appeared to have no braiding, and yet the former appearance was seen in another part of the ring. The Delta Scorpii occultation showed that the F Ring consists of at least ten individual strands. The optical depth in the thickest part of the ring is as great as in many parts of the A and B rings. It is, however, apparently restricted to a region less than 3 km wide. The thick feature has a depth of about 100 m.

The explanation is probably to be found in the presence of the two small satellites 1980 S26 and S27, which were discovered by Voyager 1 and are known as the "shepherd" satellites. These move in orbits to either side of Ring F, S26 between 2,000 and 500 km from the ring and S27 500 km closer in. Both are very small—of the order of 200 km in diameter—and both are presumably made up of ice, but their gravitational pulls are sufficient to confine the F Ring particles to a definite, narrow region. They must also be responsible for the unusual structure of the ring, though precise details are not yet known. It may be that the braiding phenomenon is either temporary, or restricted to one particular part of the ring.

The fact that the F Ring is kept in place by shepherding satellites led to the suggestion that the same might apply to the minor gaps inside the bright rings, so that these rings would contain hundreds or even thousands of tiny shepherds. Voyager 2 was put on a course that took it as close as possible to the rings in order to search for such bodies, but none came to light.

Ring F is not circular: on average it is 140,000 km from the center of Saturn, but this varies over a range of at least 400 km. Again it seems that perturbations by satellites provide a likely explanation.

Ring G

Next in order of increasing distance from Saturn is Ring G, which is 107,000 km above the cloud tops or 170,000 km from the center of Saturn's globe. Its existence was detected by experiments carried on Pioneer 11, but it was not confirmed until the Voyager passes, and even then it was seen only in forward-scattering light. It is extremely tenuous even when compared with Ring F. It lies between the orbit of Mimas and those of the two co-orbital satellites, 1980 S3 and S1, so that the average revolution period of its particles is 19.9 hr. The distance between Rings F and G is about 30,000 km. The G Ring is optically thin, but it contains sufficient material to endanger passing spacecraft. Voyager 2 passed 2,000 km from it at the time of the ring-plane crossing and it is thought that some of the tiny fragments that make up the ring collided with the spacecraft.

Ring E

Finally, there is Ring E, which extends from a distance of 147,000 km from the cloud tops at the inner edge as far as 237,000 km at the outer edge. It seems that the inner portion is brighter than the outer, and that the maximum brightness occurs at a distance of 230,000 km from the center of Saturn. The distance of Enceladus from the planet's center is 240,200 km, so that the brightest part of Ring E lies just inside Enceladus' orbit. Whether this is coincidence or not is a matter for debate. Enceladus, with its

1. Braiding
The apparently braided components of the F Ring defy the laws of orbital mechanics. The ring's complex structure includes strands that appear to cross several times, but no known law predicts what can be seen in this Voyager 1 image. It is thought that this phenomenon must result from the gravitational effects of shepherd satellites, S26 and 27.

2. Structure of the F Ring
This Voyager 2 image is an almost edge-on view of the F Ring and was taken from a range of 103,000 km. It reveals at least four faint components and one bright one. The total radial extent of these strands is about 500 km. The innermost component is the faintest: it is smooth and does not appear to interact with the other ring components.

3. Clumping material
A detail of the bright component of the F Ring reveals another interesting result from Voyager—the presence of "clumps" within the ring structure. High resolution images show up these bright clumps, some of which have very sharply defined edges. These may be individual, relatively large particles. Less well-defined patches may be concentrations of smaller particles. The orbital period of the clumps corresponds to the F Ring's period of approximately 15 hr. Again, these features may be associated with the shepherding satellites.

4

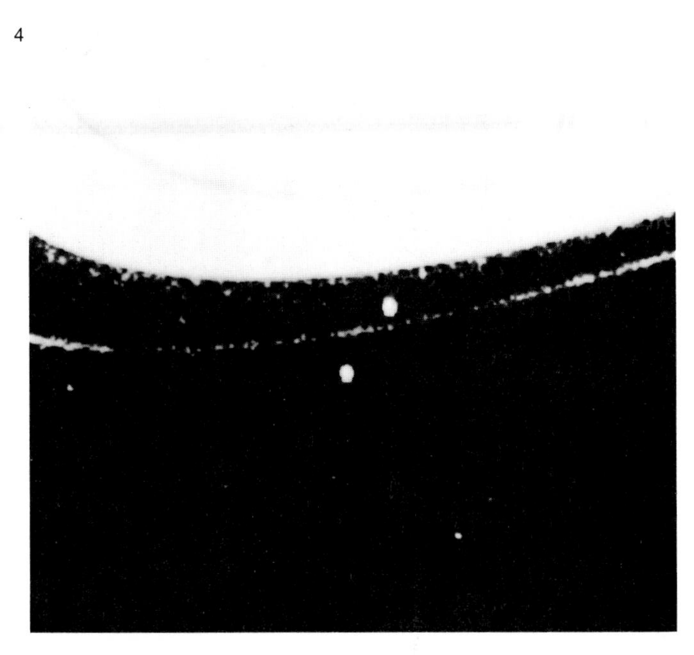

4. Guardian satellites
The F Ring has two satellite companions, 1980 S26 and 1980 S27, one either side of the ring. They may be responsible for some of the peculiar phenomena exhibited by the F Ring.

5. Shepherd satellites
Gravitational shepherding of particles within the ring system by moonlets may produce the characteristic banding and gaps. If a moonlet and two particles are in orbit around a planet (**A**), the inner particle moves faster than the moonlet, which travels faster than the outer particle. The inner particle catches up with the moonlet as it overtakes the outer particle (**B**). Gravity draws the particles closer to the moonlet just after they are neck and neck (**C**). The result is that the net gravitational pull of the moonlet lifts the outer particle's orbit and lowers the inner particle's orbit (**D**).

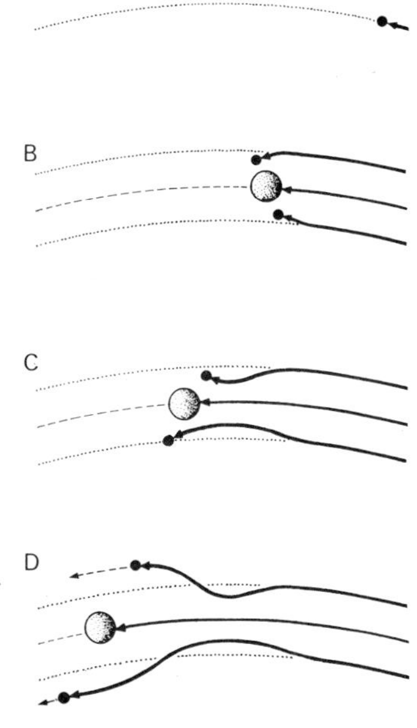

6. The G Ring
The G Ring is about 30,000 km beyond the outer edge of the A Ring, which can be seen to the right of the F Ring.

6

partially smooth surface and high albedo, may well be an active world, and there is a possibility that particles from its surface, eroded by bombarding micrometeorites, may constantly enter Ring E and replenish it. It has also been suggested that the periodical venting of material from below the surface of Enceladus, the result of tidal interactions with Dione, may also provide a source of particles for Ring E.

Are there any more rings beyond Ring E? Observations from spacecraft and ground-based measurements suggest that no further rings of comparable optical depth exist further out.

Origin of the ring system

When Saturn was the only known ringed planet, theories of the origin of the ring system were regarded as peculiar to Saturn alone. The situation today is different. Jupiter has a ring; Uranus has a whole set, though admittedly they are very different from Saturn's—the rings of Saturn are as bright as ice, whereas those of Uranus are as black as coal dust. Whether or not Neptune has a ring remains to be seen. The presence of a large satellite (Triton) moving around Neptune in a retrograde orbit, may make for unstable conditions. However, the possibility of a ring cannot be ruled out altogether.

It has been seen that Saturn's system possesses many more ringlets and divisions than was first thought. There are several examples of clumping material, representing instabilities in the system. It has been suggested that the presence of tiny moonlets could be associated with the production of many narrow features in optically thick regions and broad gaps in optically thin regions.

The first widely accepted theory of Saturn's rings was proposed more than a century ago by the French mathematician Edouard Roche. For Saturn, the Roche limit lies close to the outer edge of the main ring system. In the pre-Space Age all known planetary satellites lay outside the Roche limits for their primaries, but this is no longer true. According to Roche, a large object encountered Saturn at a fairly early stage in the evolution of the Solar System. The body may have been a large meteoroid, an asteroid, or even a moonlet that had been formed close to Saturn. Saturn's pull of gravity would create a shearing force because of differential gravitational attraction on the parts of the body closer to the planet and those parts further away. The result was that the intruder was broken up and the fragments spread around to form the ring system that exists today.

Some of Roche's conclusions have been modified in the light of more recent findings, and in particular it seems that a solid satellite less than about 100 km in diameter will not be disrupted at any distance from the planet. A large moon cannot be disrupted at a greater distance than 0.4 of the planetary radius from the surface of the body. For Saturn this would place the disruption threshold inside the inner edge of the main ring system. It is possible for two particles that differ greatly in size to resist tidal disruption at distances well within the classical limit.

An alternative suggestion is that a single large satellite moving round Saturn was broken up by collision with a wandering object such as an asteroid, so that, again, the fragments would be spread out to form a ring system. However, there is growing support for the idea that the ring particles were never part of a larger body, and that they represent the debris associated with Saturn as the planet evolved from the original solar nebula. Small grains of material can grow by condensation of vapor onto their surfaces and then by local accretion due to collisions.

Saturn's rings are probably the result of several of these processes. It is possible that the obscure ring of Jupiter is a temporary phenomenon, and the rings of Uranus are more Jovian than Saturnian in nature. As far as rings are concerned, Saturn is a planet apart. There can be no doubt that the rings will last for as long as Saturn itself.

The Satellites Introduction

Jupiter, Saturn and Uranus each possesses a system of regular satellites—bodies with nearly circular orbits in the primary planet's equatorial plane. But that of Saturn differs from the others. Jupiter has four major attendants and more than a dozen small ones, the outer members of which may be captured asteroids. Uranus has five known satellites, all between 550 and 1,800 km in diameter.

Saturn's satellite system, however, is dominated by one large body, Titan, which is larger than any other known satellite apart from Ganymede in Jupiter's system (and possibly Triton in Neptune's). The rest are much smaller. The present total of known satellites is definitely 21, and possibly 23, giving Saturn the most extensive retinue known in the Solar System.

And Saturn's satellites are not all of the same general type. All contain a high percentage of ice, but there are great dissimilarities: Dione, only slightly larger than Tethys, is much more dense; Mimas has one huge crater that is about one third the diameter of the satellite, whereas Enceladus has a surface that is partly smooth; Hyperion has an irregular shape; Iapetus has one bright and one dark hemisphere; and Phoebe has a retrograde motion and may be asteroidal in origin.

The recently discovered satellites have orbits that, while regular, display special dynamical features. They occur as ring shepherds or are co-orbital with the larger satellites. Saturn's system is the first to be studied closely, and, as a consequence, more is known about the dimensions, densities, chemical composition and possible evolution of the satellites.

1

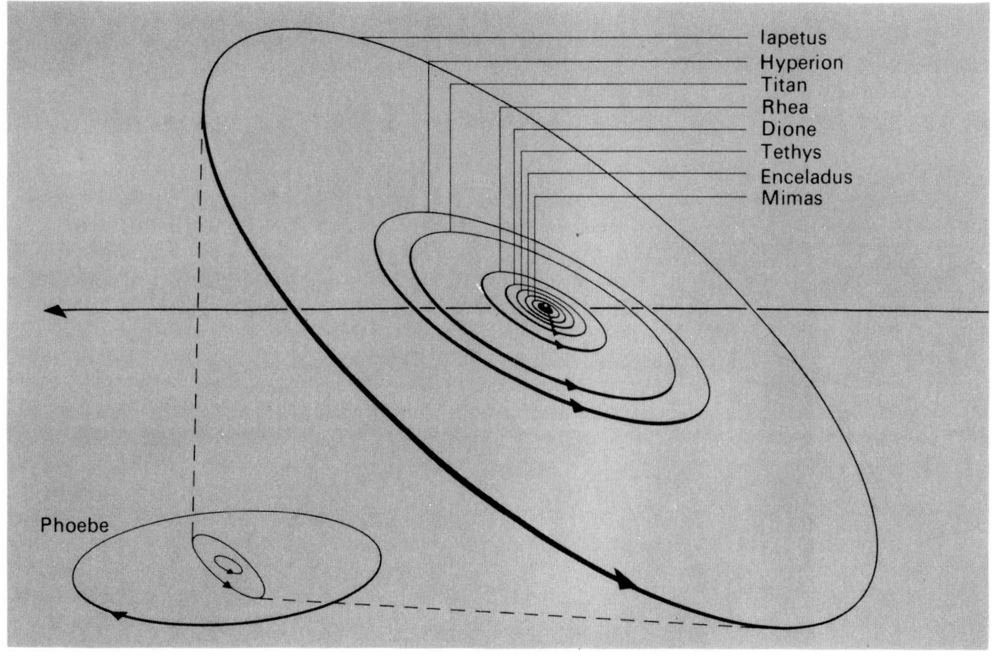

2

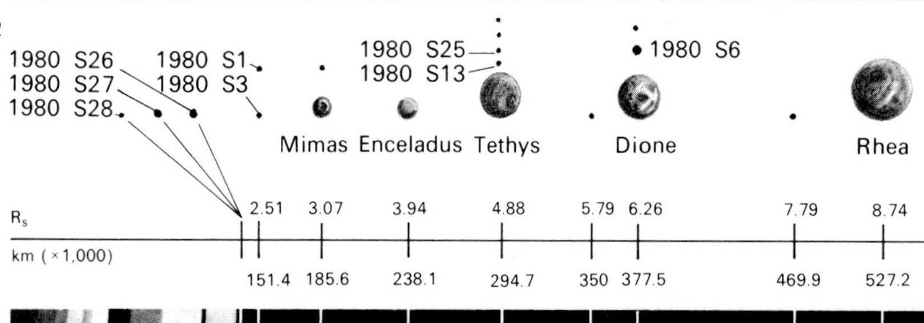

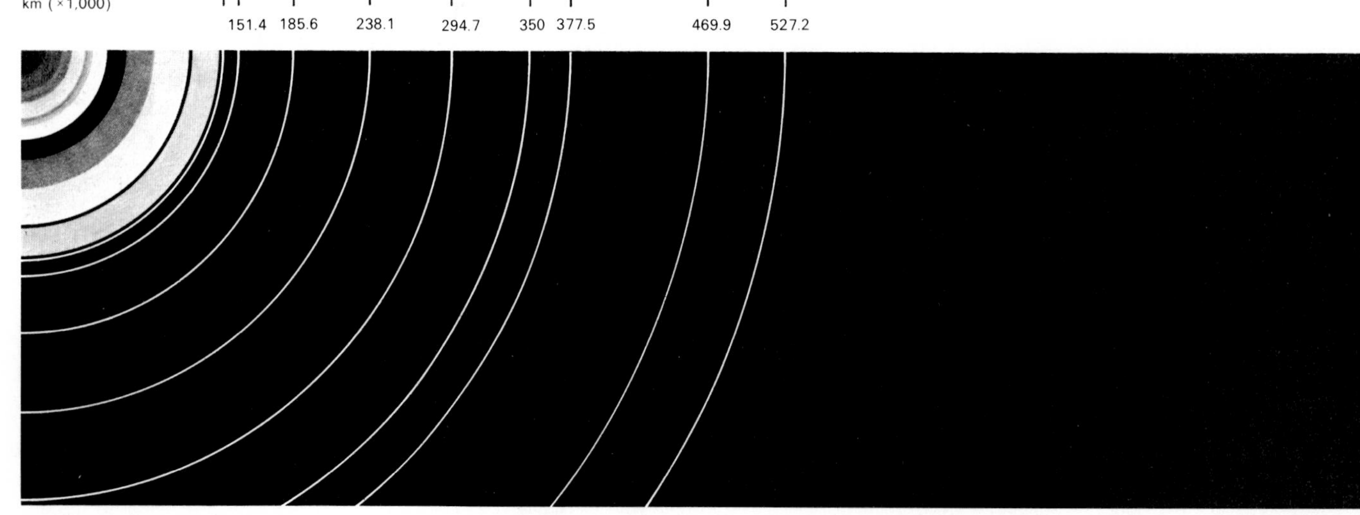

1. Satellite orbits
The orbits of Saturn's classic satellites are varied. From Mimas to Hyperion they are almost circular and lie within about one degree of the planet's equatorial plane, which is also the plane of the ring system. The orbit of Iapetus is inclined more sharply, at about 14°, while Phoebe, which lies much further from Saturn than any other satellite, has an orbital inclination of 150°. It also has a retrograde motion. Phoebe's orbit is drawn to one tenth of the scale of the orbits of Saturn's inner satellites.

2. Scale of the satellites
Most of Saturn's named moons—Mimas, Enceladus, Tethys, Dione, Rhea, Hyperion, Iapetus and Phoebe—are much smaller than the Earth's moon or Jupiter's large moons, but they are larger than most asteroids and the tiny moons of Mars, Jupiter and the recent discoveries around Saturn. Their relative sizes are shown in the diagram together with their distances (to scale) from Saturn.

The Saturnian Satellites

No.	Satellite	Discoverer	Year of discovery	Mean distance from Saturn km	Diameter km	Magni-tude	Orbital inclination degrees	Orbital eccentricity	Sidereal period days	Mean synodic period d hr min sec
	1980 S28	From		137,670	40 × 20	?	≈ 0	?	0.609166	?
	1980 S27	Voyager	1980	139,353	220	?	≈ 0	?	0.613000	?
	1980 S26	photographs		141,700	200	?	≈ 0	?	0.628541	?
	1980 S3	Fountain & Larson	1978	151,422	90 × 40	?	≈ 0	?	0.694333	?
	1980 S1	Fountain & Larson	1978	151,472	100 × 90	?	≈ 0	?	0.694667	?
	Mimas co-orbital	Synnott & Terrile	1982	≈ 185,600	10	?	≈ 1.5	≈ 0.2	≈ 0.94	?
I	Mimas	Herschel	1789	185,600	390	12.1	1.5	0.0202	0.942422	22 37 12.4
II	Enceladus	Herschel	1789	238,100	500	11.9	0.0	0.0045	1.370218	1 8 53 21.9
III	Tethys	Cassini	1684	294,700	1,050	10.4	1.1	0.000	1.887803	1 21 18 54.8
	1980 S25	Group led by Smith	1980	294,700	35	?	?	?	≈ 1.9	?
	1980 S13	Group led by Smith	1980	294,700	35	?	?	?	≈ 1.9	?
	Tethys co-orbital	Synnott	1982	≈ 294,700	≈ 15	?	≈ 0.0	≈ 0.002	≈ 1.9	?
	—	Synnott & Terrile	1982	?	?	?	?	?	?	?
	—	Synnott	1982	350,000	≈ 15	?	?	?	2.44	?
IV	Dione	Cassini	1684	377,500	1,120	10.4	0.0	0.0022	2.736916	2 17 42 9.7
	1980 S6	Lacques & Lecacheux	1980	378,060	160	?	?	?	≈ 2.7	?
	Dione co-orbital	Synnott	1982	≈ 378,000	≈ 15	?	≈ 0.3	≈ 0.001	?	?
	—	Synnott & Terrile	1982	469,900	?	?	?	?	?	?
V	Rhea	Cassini	1672	527,200	1,530	9.8	0.3	0.0010	4.517503	4 12 27 56.2
VI	Titan	Huygens	1655	1,221,600	5,140	8.3	0.3	0.0292	15.945448	15 23 15 31.5
VII	Hyperion	Bond	1848	1,483,000	400 × 250 × 240	14.2	0.6	0.1042	21.277657	21 7 39 5.7
VIII	Iapetus	Cassini	1671	3,560,000	1,440	10–12	14.7	0.0283	79.33085	79 22 4 59
IX	Phoebe	Pickering	1898	12,950,000	160	16	150	0.1633	550.337	523 13 — —

Properties of the largest Saturnian satellites

Satellite	Mean orbital radius km	Mass kg	Density g/cm³	Albedo
Mimas	185,600	3.76×19^{19}	1.44 ± 0.18	0.7
Enceladus	238,100	(7.40×10^{19})	1.16 ± 0.55	1.0
Tethys	294,700	6.26×10^{20}	1.21 ± 0.16	0.8
Dione	377,500	1.05×10^{21}	1.43 ± 0.06	0.5
Rhea	527,200	(2.28×10^{21})	1.33 ± 0.09	0.6
Titan	1,221,600	1.36×10^{23}	1.88 ± 0.01	0.2
Hyperion	1,483,000	(1.10×10^{20})	?	0.3
Iapetus	3,560,100	(1.93×10^{21})	1.16 ± 0.09	0.5, 0.05

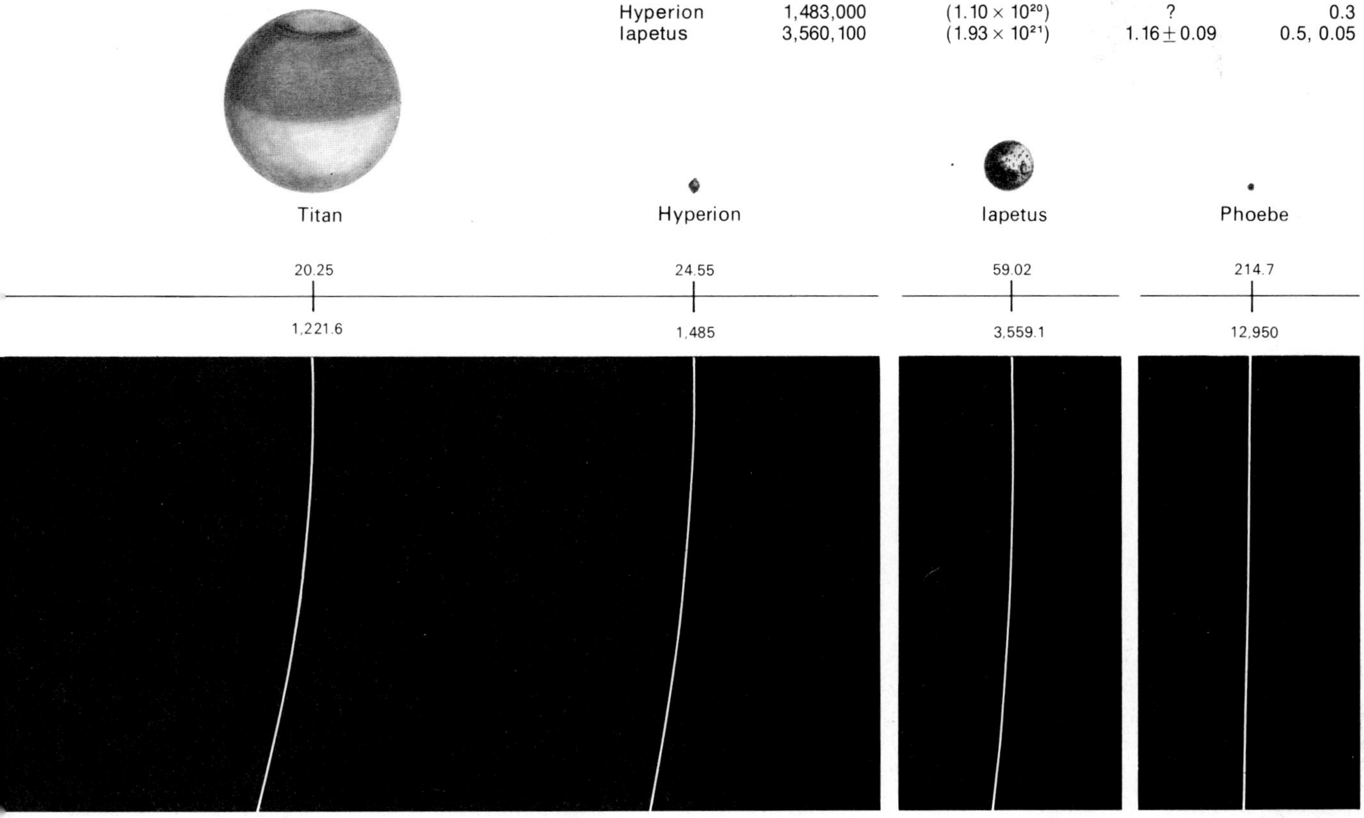

Titan

20.25

1,221.6

Hyperion

24.55

1,485

Iapetus

59.02

3,559.1

Phoebe

214.7

12,950

The Satellites History

Titan, the largest of Saturn's satellites, was also the first to be discovered. It was recorded on 25 March 1655 by Christiaan Huygens, and it took him little time to decide that it really was a satellite rather than a star; it moved with Saturn, and Huygens found that its revolution period was 16 days. It is an easy object, with a magnitude just below 8.

Cassini's discoveries

The next discoveries were made by G. D. Cassini. Like Huygens, Cassini used small-aperture, long-focus refractors, with lenses made by the best instrument-makers of the period such as Campani and Divini. In 1671, using one of Campani's telescopes, he discovered another satellite, much further away from Saturn, and calculated that its revolution period was a little less than 80 days; it was named Iapetus. The true revolution period is 79.3 days, so Cassini's estimate was very accurate. He also found, again correctly, that the orbit is appreciably inclined to the plane of Saturn's equator; in fact by 14.7°. Moreover, he observed that the brilliancy of Iapetus was not constant. When west of the planet it was an easy object, but when east of Saturn it became too faint for him to see with his telescope. He therefore concluded that "one part of his surface is not so capable of reflecting to us the light of the Sun which maketh it visible, as the other part is".

This also was correct, though final proof was not obtained until the Voyager missions. It followed from this observation that the rotation period was captured or synchronous; that is, Iapetus keeps the same hemisphere turned permanently towards Saturn, as the Moon does to Earth. Tidal friction is responsible for this, and all major planetary satellites behave in the same way.

In 1672, when he announced the variability of Iapetus, Cassini discovered another satellite, now known as Rhea, which is of approximately the same magnitude as Iapetus at its best. Then, in March 1684, he made new studies, using an unwieldy aerial telescope with a focal length of more than 40 meters, and discovered two rather fainter satellites, Dione and Tethys. Their magnitudes are between 10 and 10.5, so that they again are not difficult objects; and both have orbits closer to Saturn than that of Titan.

By the end of the seventeenth century five satellites were known, all of which had nearly circular orbits, and of which only Iapetus had an orbit appreciably inclined to Saturn's equatorial plane.

Herschel's discoveries

On 19 August 1787 William Herschel, using his 6-meter reflector (aperture 46 centimeters), suspected a new inner satellite, but did not follow up his observation until August 1789, by which time he had completed his 12-meter reflector (aperture 122 centimeters). Herschel soon confirmed his original suspicion, and discovered a seventh satellite still nearer to the planet; these new objects were named Enceladus and Mimas.

The discovery of Hyperion

The first major American observatory was at Harvard. From there studies of Saturn were carried out in 1848, and G. P. Bond detected "a star of the 17th magnitude in the plane of Saturn's ring, between Titan and Japetus" ("Japetus" was the old spelling of "Iapetus"). It was soon confirmed, both by G. P. Bond and his father, W. C. Bond, Director of the Observatory. By mid-October they had found that the revolution period was a little over 21 days, and that the satellite moved practically in the plane of Saturn's equator.

Before the news reached England, William Lassell, using his reflector of 6 meters focal length, had found the satellite. His name for it—Hyperion—was accepted. Lassell rightly concluded that Hyperion is intrinsically fainter than Mimas, but was easier to see because it is so much further from the planet. He also suggested that its brightness was decidedly variable. The modern estimate of its mean magnitude is 14.2, considerably brighter than Bond's original

1

1. Christiaan Huygens 1629–95
Huygens discovered Titan in 1655 with a magnification of 50 on his small-aperture refractor of long focal length. The observation was soon confirmed by other observers. Huygens failed to find any other satellites around Saturn and he correctly concluded that if any other moons existed they must be fainter than Titan.

2. Herschel's giant telescope
This 12-meter reflector was by far the largest telescope made at that time, and it was not surpassed until Lord Rosse's 122 cm reflector in 1845. Herschel's telescope was, however, clumsy and unwieldy to use and his discovery of Enceladus and Mimas with it was perhaps its greatest contribution.

2

3

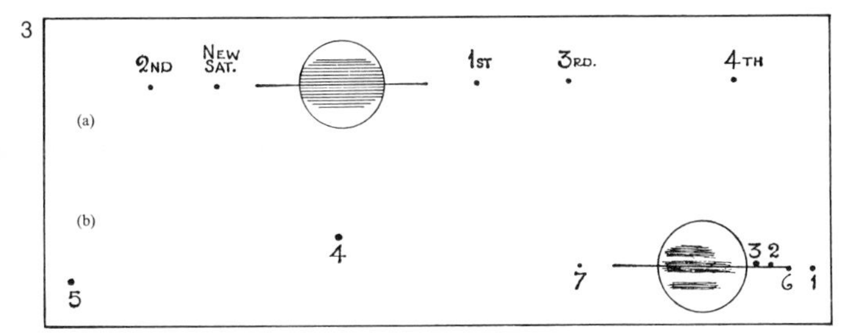

3. Herschel's sketches
In 1789 Herschel illustrated the alignment of Saturn's satellites in August (top) and then in October (bottom) to show his new discoveries—Enceladus (6) and Mimas (7). The other satellites drawn are Tethys (1), Dione (2), Rhea (3), Titan (4) and finally Iapetus (5).

4. Drawings of Titan
These remarkable drawings were made by B. Lyot, A. Dollfus and their colleagues at the Pic du Midi Observatory. Observations and drawings of Titan's tiny disc were made independently between 1943 and 1950. These drawings represent features that were detected by several observers.

4

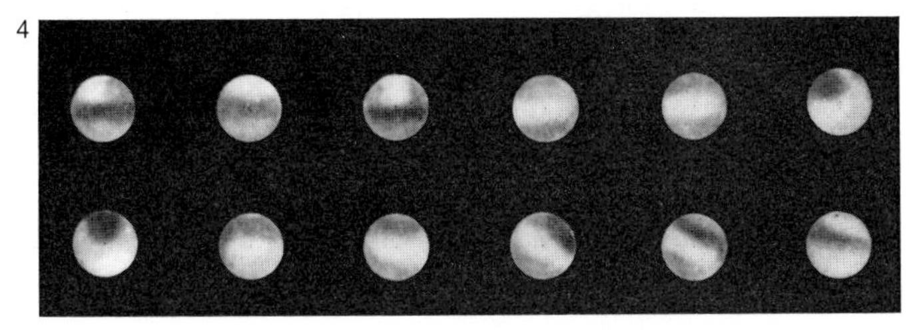

5

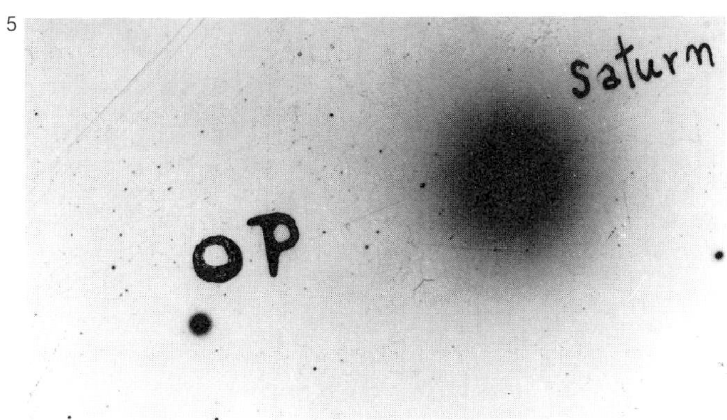

5. Discovery of Phoebe
Pickering's discovery of Phoebe from photographs was the first of its kind. Between 16 and 18 August 1898, he exposed four plates for two hours each. The plates were compared and an object—not a star—was found to have moved. Saturn's ninth satellite had been found.

6. The "discovery" of Janus
From photographs taken at Pic du Midi Observatory in December 1966, Dollfus believed that he had found yet another satellite, which he named Janus. He estimated its period to be 0.815 days and its diameter 300 km, but Saturn's rings began to open out and any further search was impossible.

6

value. Changes in magnitude are much less easy to detect than with Iapetus because Hyperion is so much fainter, but the Voyager 2 pictures have shown that the satellite is irregular in shape, so that the magnitude may be expected to vary to some extent, and this could explain Lassell's findings.

Pickering's results

In 1898 W. H. Pickering, a foremost American planetary observer, began a search for new satellites of Saturn from Arequipa in Peru, the southern station of the Harvard College Observatory. The method used was photographic; the telescope used had an aperture of 61 centimeters. Four plates were exposed for two hours each, and they revealed more than 400,000 stars between them. A new satellite was found to be orbiting well beyond Iapetus and Pickering named it Phoebe. For some time its existence was questioned, but in 1904 E. E. Barnard confirmed it from the Yerkes Observatory. It was found to have retrograde motion and to be smaller than any of the previously discovered satellites. Despite its great elongation from Saturn, which can exceed 34 minutes of arc, it is difficult to see visually. The magnitude is about 16.

Shortly afterwards, Pickering believed that he had discovered another satellite, moving around Saturn at a distance of 1,460,000 km in a period of 20.85 days; this would have placed it between the orbits of Titan and Hyperion. The photographs concerned had been taken in 1899, but Saturn was then in a very rich region of the Milky Way, and positive identification was difficult. Before Pickering's claim could be verified, the satellite apparently vanished! It was named—Themis—and elements of its orbit continued to be published in some astronomical almanacs (such as the *Connaissance des Temps* in France) as recently as 1960, but it has never been confirmed, and it now seems certain that Themis does not exist. Pickering was misled either by a star or, more probably, by an asteroid.

Janus

When the rings of Saturn are wide open, the fainter satellites, and particularly those close to the planet, are difficult to observe. When the rings are edgewise-on, as happens every 15 to 17 years, conditions are much more favorable. The rings were so placed in 1966, and careful searches were made by A. Dollfus at the high-altitude Pic du Midi Observatory in the Pyrenees. His photographs indicated the existence of an inner satellite, whose distance from Saturn was given as 169,000 km. A few confirmatory observations were obtained, but many observers were sceptical about the existence of the new satellite, named Janus. From Earth, no confirmation could be expected before the next edgewise presentation of the rings, in 1980. Meanwhile the Voyager missions had been planned, and results from them have shown that there is no such satellite moving in the orbit given for Janus. There are several small inner satellites, and one of these may have been observed in 1966, but the name Janus has been deleted from the official lists of Saturn's satellite family.

New inner satellites

The Pioneer and Voyager results have shown that there are several small satellites moving within the orbit of Mimas (*see* pages 60–61). There are also small bodies moving in the same orbits as the larger satellites. Dione has two co-orbitals, one moving 60° ahead of Dione and the other 60° behind at the so-called Lagrangian point. Since they always keep in virtually the same position relative to Dione, there is no fear of a collision. Ground-based observations have also shown that there are two small satellites moving in the same orbit as Tethys. Other co-orbitals may be associated with Mimas and Enceladus. There is every possibility that other minor satellites in the Saturnian system remain to be discovered.

The Satellites Observations

Saturn's greater distance from Earth means there can be no observations comparable with those of Jupiter's Galilean satellites, which show obvious discs even in a small telescope. Unsuccessful efforts were made to record surface details of Titan, the only large satellite in Saturn's retinue, but the surface is permanently concealed below the atmospheric "smog". Even Voyager 1 showed virtually nothing. All that could be said from ground-based observations was that Titan appeared decidely orange or yellowish in color. Before the Voyager missions nothing was known of the surface features of the other satellites.

There was, however, an important development in 1943–44, when G. P. Kuiper studied Titan spectroscopically, and reported the existence of an atmosphere that was assumed to be composed chiefly of methane. In 1977, L. Trafton reported the existence of molecular hydrogen. Voyager 1 showed the atmosphere of Titan to be mainly nitrogen, with some methane, and a ground pressure 1.6 times that of the Earth at sea-level.

Transits, eclipses and occultations

These phenomena, so familiar to all observers of the Galilean satellites of Jupiter, are much less easy to follow in Saturn's system. Transits and shadow transits of Titan can be seen and the phenomena may also be detected with Rhea, Dione and Tethys, although telescopes of considerable aperture are needed. The shadow of Rhea on Saturn's disc has, however, been glimpsed with a reflector as small as 15 cm aperture.

Eclipses and occultations of the brighter satellites (Titan, Rhea, Iapetus, Dione and Tethys) are not difficult to observe. Those of Iapetus are the most interesting, because it is so much further from Saturn and because it has an orbit appreciably inclined to Saturn's equatorial plane. An early observation was made in 1889 by E. E. Barnard, with a 30 cm refractor; Iapetus was eclipsed by Ring C. The last opportunities to date occurred in 1977 and 1978, and P. Doherty, using a 41 cm reflector, found that the satellite was never lost while in the shadow of Ring A; and a slight brightening was produced by the Encke Division.

Mutual phenomena are very rare indeed, but can be observed occasionally: thus, in 1921 W. H. Pickering described occultations of Rhea by Titan, and of Dione by Rhea. The only really well-observed eclipse of one satellite by the shadow of another was that of 8 April 1921, when Rhea passed into the shadow of Titan for more than half an hour.

The Voyagers

During the Voyager missions, great attention was paid to the observation of satellites. Voyager 1 made close-range surveys of Titan, Rhea, Dione and Mimas. Titan was regarded with special interest. Although the surface was found to be masked by clouds, valuable measurements were obtained. Voyager 2 surveyed Iapetus, Hyperion, Tethys and Enceladus.

Dione and Rhea were surveyed by both Voyagers. Voyager 2's closest approach to Rhea was over the satellite's north pole, and close-range pictures were obtained, revealing that the north polar region contained zones with different albedoes. Voyager 1 obtained preliminary images of Tethys, including the first indications of the huge trough now known as Ithaca Chasma, and Voyager 2 extended the survey to a much wider area of the surface. Voyager 2 uncovered the most interesting characteristics of Enceladus—indications that the surface may still be active. It became clear from Voyager 1 that the trailing hemisphere of Iapetus was bright and the leading hemisphere dark; the main charting, however, was done by Voyager 2, and this was also the case with the irregularly shaped Hyperion. Phoebe, unfortunately, was not within useful range of either Voyager, and all that can be said as yet is that the surface is darkish, and the rotation period is so far unique in Saturn's family in being non-synchronous.

1A

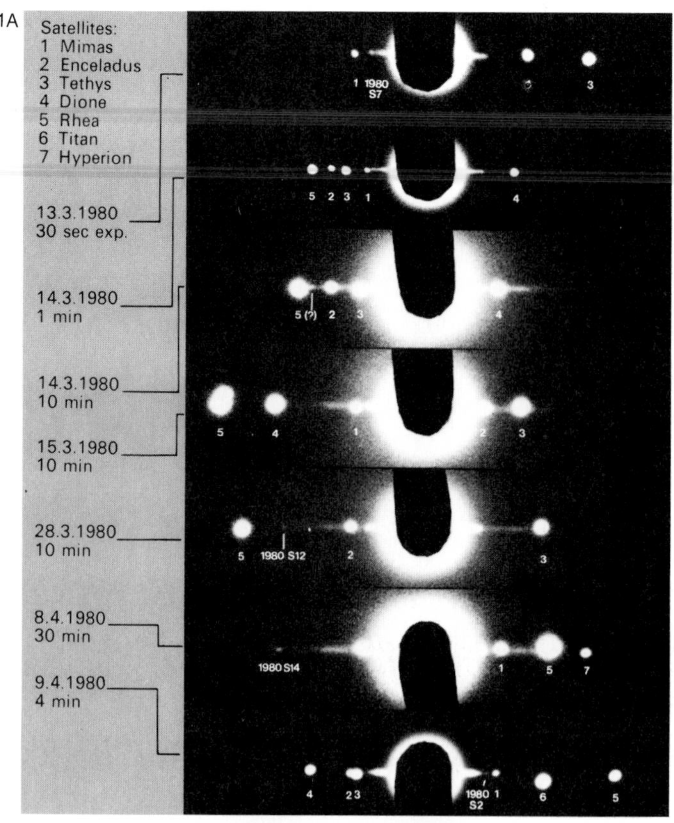

B

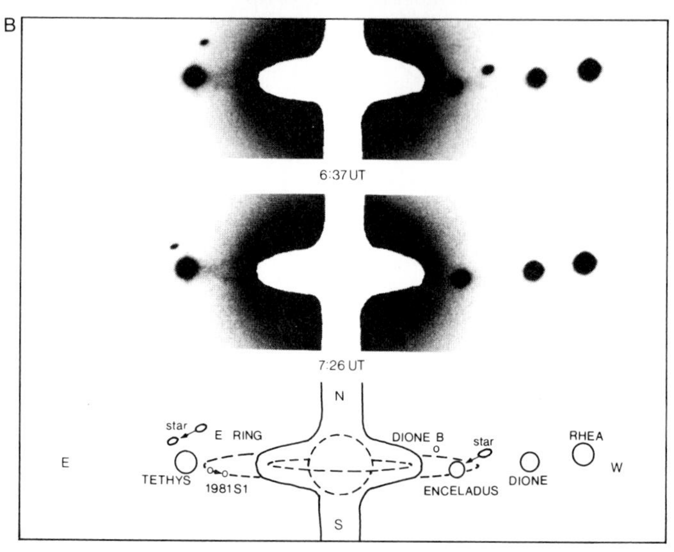

1. Saturn ring-plane crossing
A series of observations were made during March (**A**) and April (**B**) in 1980 from the Catalina Observatory. They show the satellites in the equatorial plane of Saturn—Mimas (1), Enceladus (2), Tethys (3), Dione (4), Rhea (5), Titan (6) and Hyperion (7).

2. Satellite phenomena
The various configurations of the satellites provide interest for observers. The more common examples include shadow transits (**A**) as a satellite passes in front of Saturn, and eclipses (**B**), when the satellite passes behind the planet and in its shadow.

2A

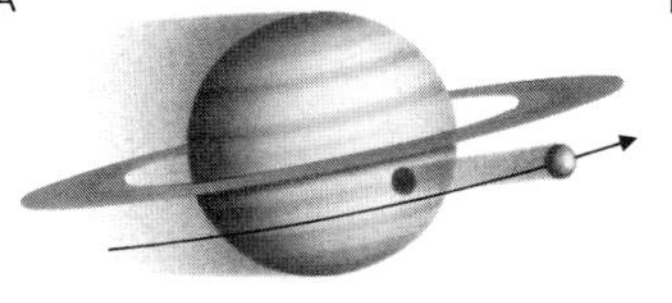

B

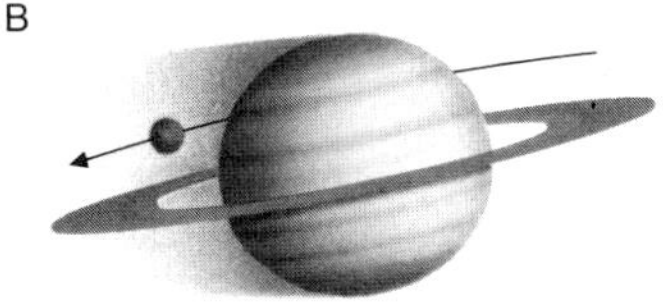

3. Voyager 1 flyby of Saturn
The equatorial view of Voyager 1's trajectory shows the spaceprobe's path in relation to the planet's equatorial plane, and hence the ring system and the orbital plane of the inner satellites. The polar view shows the trajectory in relation to the rings and satellite orbits. The satellites are not drawn to scale and are placed at the point of Voyager's closest approach in light. The smaller diagrams show the flight path as seen from Earth and the trajectory with respect to the south pole and rings.

Closest approach distances (V1)

Satellite	km
1980 S28	219,000
1980 S27	300,000
1980 S26	270,000
1980 S3	121,000
1980 S1	297,000
Mimas	88,440
Enceladus	202,040
Tethys	415,670
1980 S25	237,332
1980 S13	432,295
1980 S6	230,000
Dione	161,520
Rhea	73,980
Titan	6,490
Hyperion	880,440
Iapetus	2,470,000
Phoebe	13,537,000

4. Voyager 2 flyby of Saturn
Similar views—equatorial and polar—show the trajectory of Voyager 2 nine months later. The spacecraft's flight path was considerably revised from original plans in order to take another look at the unexpected and unexplained phenomena that were observed by Voyager 1. Voyager 2 was reprogrammed in flight so that the encounter with Saturn would further explore the results from the first mission. The paths of both Voyager flybys were calculated in each case to include occultations of the Earth, the Sun or stars by Saturn, various satellites and the rings. The small diagram shows Voyager 2's encounter with the outer satellites—Iapetus, Hyperion and Phoebe.

Closest approach distances (V2)

Satellite	km
1980 S28	287,170
1980 S27	246,590
1980 S26	107,000
1980 S3	147,010
1980 S1	222,760
Mimas	309,990
Enceladus	87,140
Tethys	93,000
1980 S25	284,396
1980 S13	153,518
1980 S6	318,200
Dione	502,250
Rhea	645,280
Titan	665,960
Hyperion	470,840
Iapetus	909,070
Phoebe	1,473,000

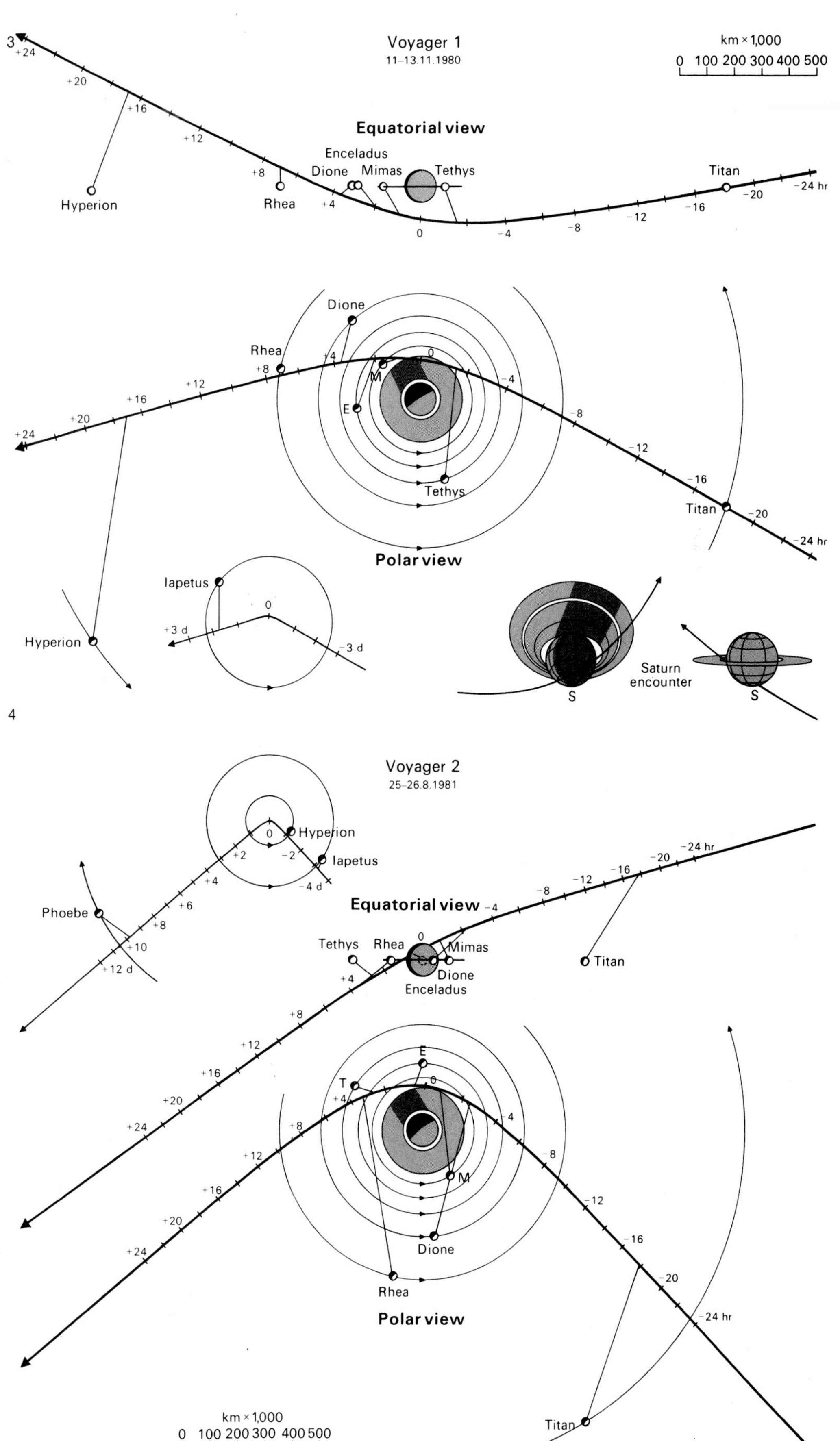

The Small Satellites

The number of known members of Saturn's system of satellites has increased considerably in recent years. Before Pioneer 11 and the Voyager flights nine satellites were known; now we know there are perhaps 23, of which five small ones orbit closer to Saturn than Mimas. They have not yet been named, and are known only by their temporary designations.

First there is 1980 S28, which moves just beyond the edge of the A Ring. It was discovered on the Voyager 1 photographs, and is an elongated object; its maximum diameter may be as much as 40 km, and it has an orbital period of 14.5 hr. Its presence may ensure that the outer edge of the A Ring is comparatively well defined.

The next two satellites, 1980 S27 and 1980 S26, are considerably larger, with diameters of at least 200 km. They move in orbits lying to either side of Ring F, and they are known as shepherd satellites because their gravitational forces keep the particles of Ring F in a stable orbit.

Further out still, 151,500 km from Saturn and 10,000 km beyond the F Ring—approximately midway between rings F and G—are two very interesting satellites, 1980 S3 and 1980 S1. Each is irregular in shape; the former measures about 90 by 40 km, and its companion 100 by 90 km. They give every indication of being fragments of a former single body which broke up, and this is made even more probable by the fact that their orbits are practically identical. In fact, the distance between their orbits is less than the sum of the diameters of the bodies, which leads to an extraordinary state of affairs which has been likened to a game of cosmic musical chairs. At present (1982) 1980 S3 is slightly closer to the planet, and its revolution period is 16.664 hr as against 16.672 for 1980 S1. This means that it will slowly catch up with its companion, but as they approach each other there will be a mutual interaction such that the inner, faster-moving body will be slightly slowed down while the outer, slower one will be speeded up. The end result is that the two satellites interchange orbits. The interval between two successive encounters is four years. Collision cannot take place, otherwise the two bodies would not continue to exist independently.

In February 1982 S. P. Synnott of the Jet Propulsion Laboratory, Pasadena, announced that the Voyager 2 data had shown the existence of another small inner satellite, moving at about the same distance from Saturn as Mimas and perhaps co-orbital with it, but this has yet to be confirmed. Its diameter is estimated at 10 km.

Further out are more co-orbital situations. Tethys has two, 1980 S13 moving 60° ahead of it and 1980 S25 60° behind; Synnott suspects a third. This is what is termed a Lagrangian situation; it was first described by the French mathematician Lagrange in 1772. A body moving either 60° ahead of or 60° behind an orbiting satellite will be stable, and will be in no danger of being pulled into destruction. These so-called Lagrangian points had already been proved by the groups of asteroids known as the Trojans, which move in the same orbit about the Sun as Jupiter, oscillating around their stable Lagrangian positions and always keeping at a respectful 60° or so from Jupiter.

The satellite 1980 S6 is often called Dione B, and was the first of the co-orbitals to be detected. It was discovered by the French astronomers P. Lacques and J. Lecacheux, not from the Voyager results but by telescopic observation in 1980, when the ring system was edgewise-on as seen from Earth, and so conditions were ideal for observing small, faint satellites close to the planet. Dione B is about 160 km in diameter, which makes it about equal in size to Phoebe. Nothing is known about its surface, but it is probably icy in nature. Synnott has identified a second Dione co-orbital.

Two other satellites reported by Synnott include one between the orbits of Tethys and Dione at a distance of 350,000 km from Saturn in a period of 2.44 days; and one further out, between the orbits of Dione and Rhea, at 470,000 km from Saturn and with a period of 3.8 days. These are probably in the order of 15 to 20 km in diameter. Many faint additional satellites may yet be discovered.

1

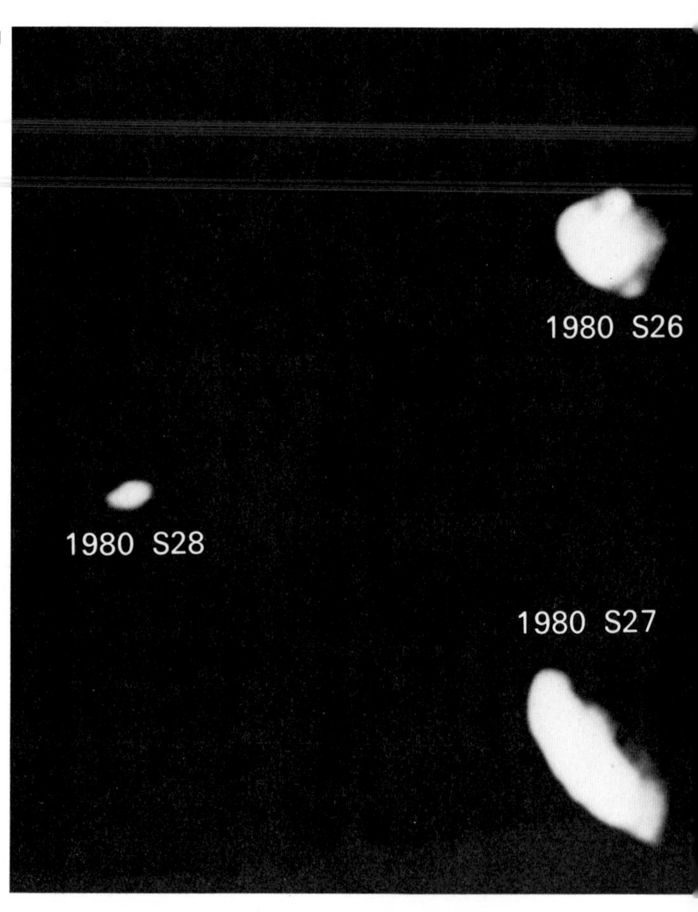

1. Minor satellites
Eight minor satellites in the inner part of the Saturnian system were observed by Voyagers 1 and 2. All eight either "guard" a ring or "share" an orbit. From the left they are 1980 S28 (A Ring shepherd); 1980 S26 and 1980 S27 (F Ring shepherds); 1980 S1 and 1980 S3 (the co-orbitals); 1980 S13 and 1980 S25 (Tethys Lagrangians); and 1980 S6 (Dione B). In this montage the images are correct to their relative scale. Several of them clearly show extensive cratering.

2

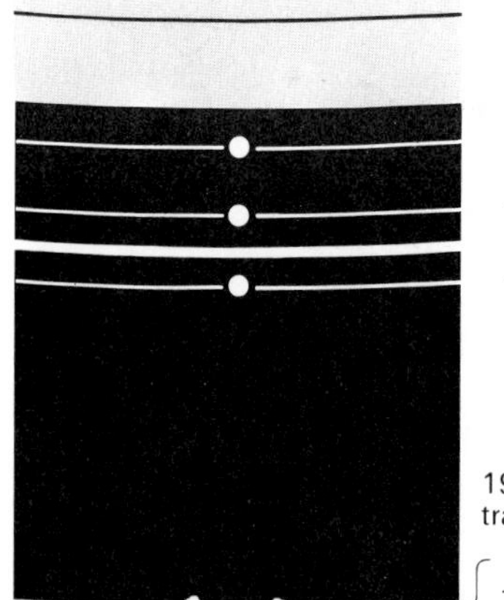

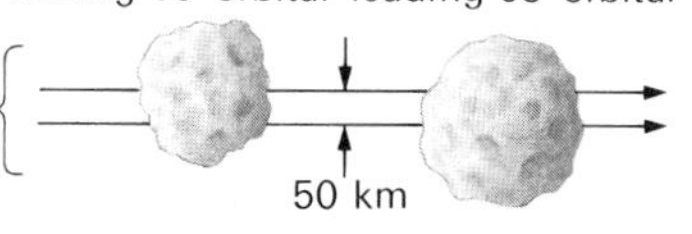

2. Positions of inner satellites
The positions of the five inner members of Saturn's recently discovered smaller satellites are shown with respect to the ring system. The A Ring shepherd (1980 S28) is the smallest satellite of Saturn yet to be discovered. The F Ring shepherds (1980 S27 and S26) effectively hold the F Ring in position. Finally, the co-orbital moons (1980 S1 and S3) may well be the fractured halves of what was once a single satellite which was disrupted in the remote past.

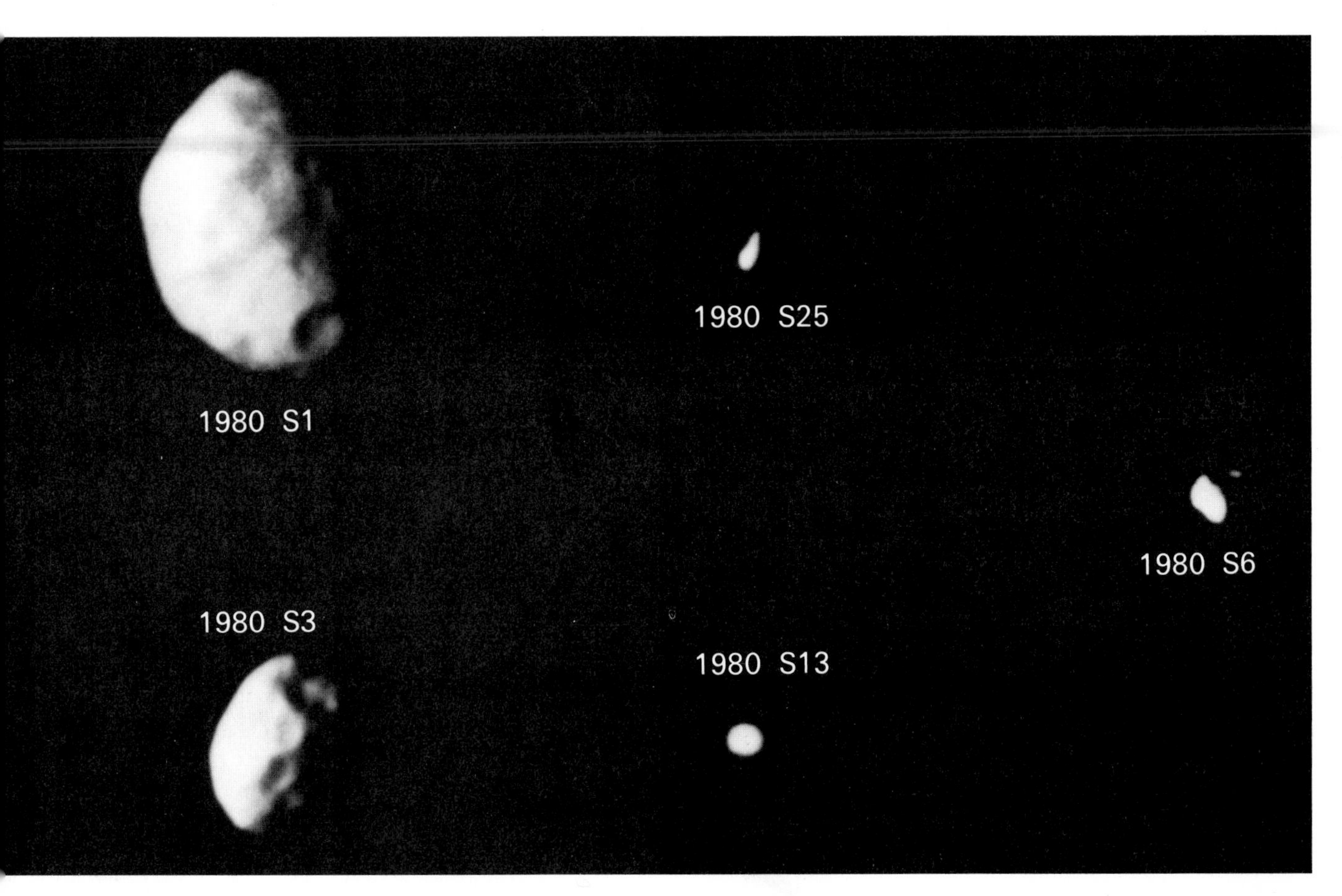

3. Transit of 1980 S3
Two Voyager 1 pictures of the eleventh moon of Saturn's system, 1980 S3, were taken 13 minutes apart and show a narrow shadow moving across the satellite. It is believed that the shadow was of the F Ring, a few thousand km away from the satellite. 1980 S3 is a trailing co-orbital and has an irregular shape and a cratered surface. These pictures show the south polar region of the moon and they were taken at a range of 177,000 km with different filters and exposure times.

4. Inner F shepherd
Proof of the F Ring's guardian satellite—1980 S27—is in this dramatic photograph of satellite and ring. The presence of moons within the ring system like this was an interesting find. Some scientists believe there may be more as yet undiscovered, but dominating the dynamics of the rings. These irregularly shaped bodies lie well within the Roche limit: ring particles may still be accreting or the satellites could be torn apart in the future, adding material to the rings.

5. Orbit sharing
Co-orbital satellites have differential orbital periods which result in gravitational interactions between the two bodies every so often. The innermost co-orbital, with a lesser orbital period, gradually approaches its companion (**A**) and, in so doing, extracts orbital momentum from it. The added momentum raises it to a higher orbit (**B**), slowing it down relative to its companion, which is dropped to a lower, faster orbit (**C**). The two bodies have, therefore, exchanged orbits (**D**). This is what is believed to happen with Saturn's co-orbitals 1980 S3 and 1980 S1, which lie roughly halfway between Rings F and G. Smaller bodies are also known to co-orbit with larger satellites. They are found at what is known as the Lagrangian points—60° to either side of the larger satellite. These are points of stability so that the smaller bodies survive and are not torn apart by gravitational forces. Tethys has two such companions—1980 S25 and S13. There may well be other such interactions elsewhere.

3

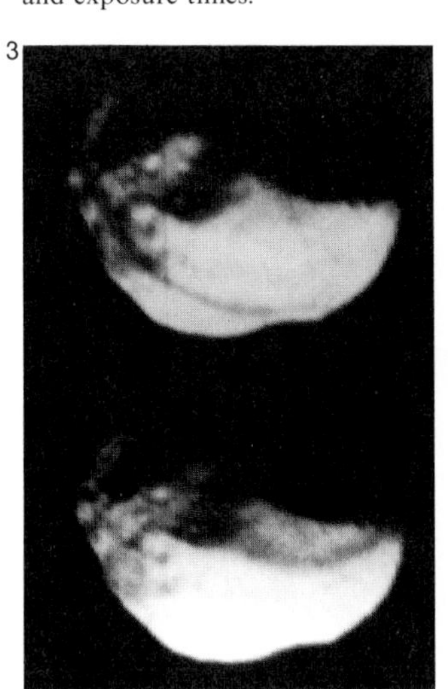

4

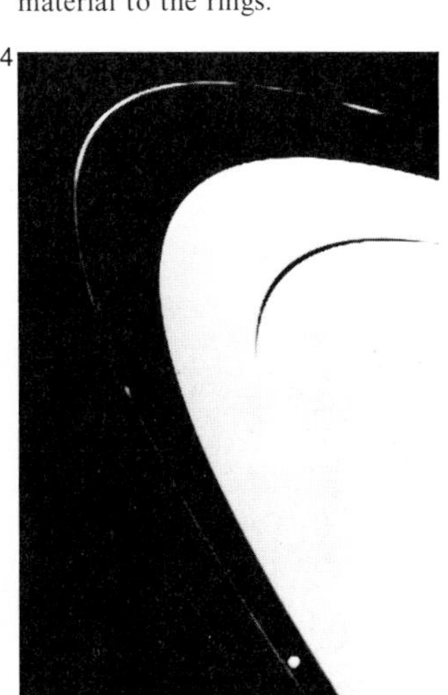

5

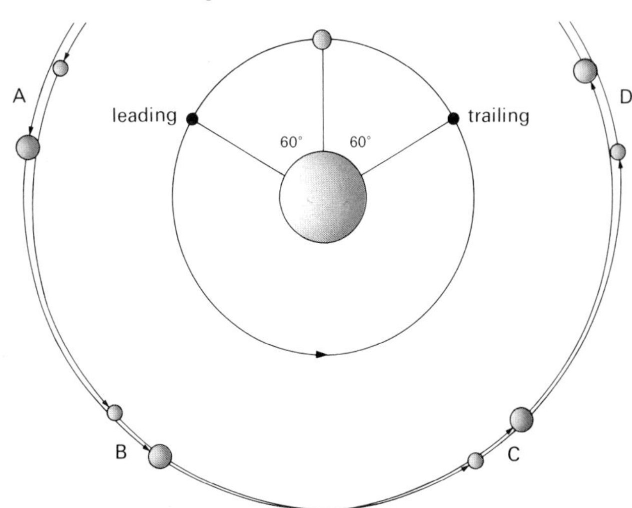

Mimas

Mimas, the innermost of Saturn's classic satellites, is a faint telescopic object. It has a diameter of 390 km—a value reliable to within a range of only 10 km. Practically all our knowledge of its surface has been drawn from the observations made from Voyager 1, which approached to within a distance of 88,400 km; Voyager 2 passed by at a minimum distance of more than 300,000 km, which was too far away to obtain large-scale images. Like all the satellites apart from Phoebe, Mimas has a synchronous or captured rotation, so it keeps the same hemisphere turned towards Saturn all the time; the revolution period and the rotation period are the same. Most of the surface was surveyed from Voyager 1, although the north polar regions were inaccessible.

Craters

The most striking feature is a giant crater, Herschel, 130 km in diameter, nearly centered on the leading hemisphere. The walls rise on average to a height of 5 km above the floor, which in parts is 10 km deep. An enormous central peak, 20 km by 30 km at its base, rises 6 km from the crater floor. This feature may have been generated by a rebound of the floor under the extremely weak gravity of Mimas. This peak is almost exactly on Mimas' equator. The crater floor includes considerable detail: adjoining it is a valley and numerous smaller craters.

The diameter of the crater is one third that of Mimas itself. This raises interesting questions: the density of Mimas is only 1.2 times that of water, and for a body of its size the diameter of the crater is probably near the maximum that can be produced by meteoritic impact without breaking up the satellite.

All the other craters are a good deal smaller than this giant feature, and only a few craters larger than 50 km have been observed. Smaller craters, however, are abundant and in general more uniformly distributed, and many of them with diameters greater than 20 km have pronounced central peaks. The region from 40°W to 260°W has very few craters in the range 20–50 km and it may have a younger surface on which later processes have obliterated preexisting large craters. The number of such craters increases substantially from 260°W to the giant crater at 100°W.

Most of the craters on Mimas are approximately bowl shaped and much deeper than craters of comparable size on the Moon or the Galilean satellites. The greater depths are probably due to the extremely low gravity field of Mimas. Craters are frequently superimposed upon other craters, while many of the older features are strongly degraded. No ray craters have been seen on the surface of Mimas, perhaps as a result of the intrinsic brightness of the mature surface itself. The apparently uniform distribution of craters with diameters less than 30 km may indicate an equilibrium in the production and destruction of the features. The giant crater is relatively unmodified by superimposed craters, and may well be younger than the rest of the surface of the satellite.

There are valleys, too, on Mimas and the surface of the satellite is scored by grooves, which extend to 90 km in length, and are generally 10 km wide and 1 to 2 km deep. The most conspicuous grooves trend northwest and west-northwest, rather like the "grid system" on the Moon. Some are straight and may have formed over deep-seated fractures or fracture systems. The less-regular systems may actually consist of chains of coalesced craters. The grooves may have been produced when the giant crater was formed. Alternatively they may have been developed by tidal interactions as the body cooled, and then froze. A few local hills are evident in the trailing hemisphere which are mostly 5 to 10 km across and less than 1 km high.

The surface of the satellite is icy, and there is every reason to suppose that ice makes up much of the entire body. The overall surface of Mimas is consistent with a surface coated with water frost and bearing scars, which some believe are the result of bombardment from exterior bodies.

1

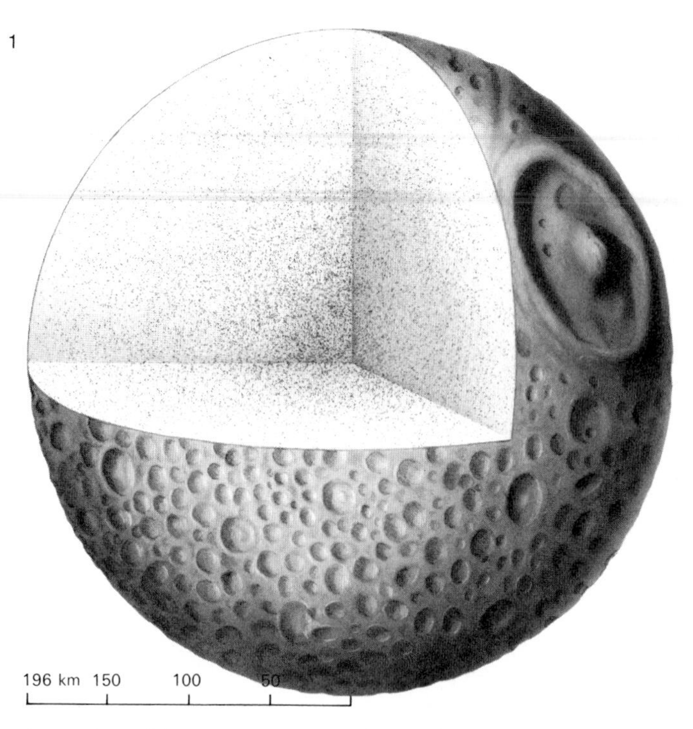

1. Interior of Mimas
Mimas is a regular icy satellite of low density. Its surface is, at least in part, covered with water ice, but little is known of the internal structure other than that its composition is at least 60 percent ice, condensed out from the initial Saturnian nebula. The remaining "rocky" material may result from accretion and bombardment from external sources.

2. Terminator
The terminator of Mimas shows a heavily cratered surface, which may represent a record of the bombardment that occurred in the Solar System during its early history, some 4 billion years ago. Some of the small craters visible in this photograph, taken at a range of 129,000 km, are as tiny as 2 km in diameter. Craters of this size are abundant on Mimas.

2

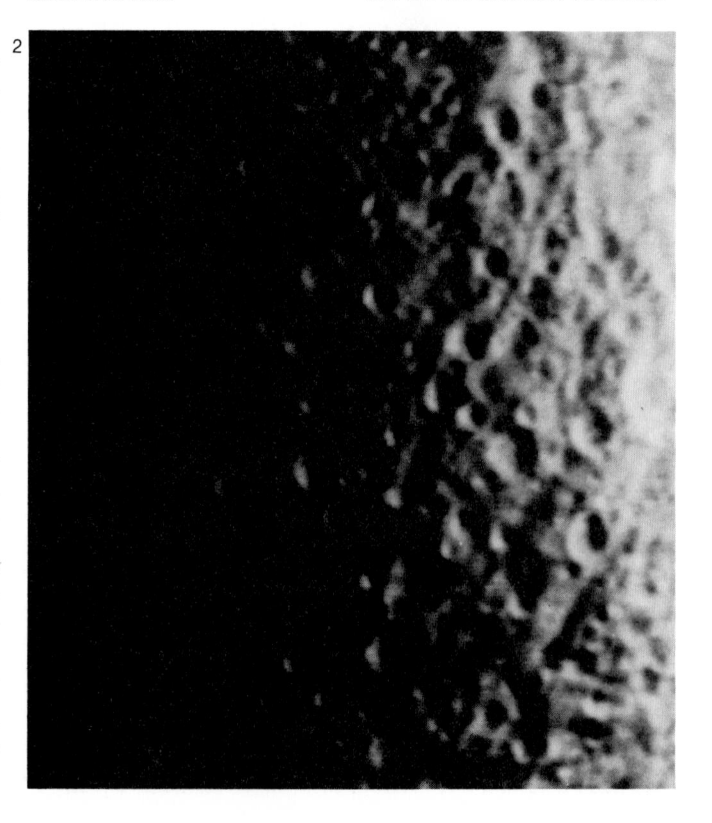

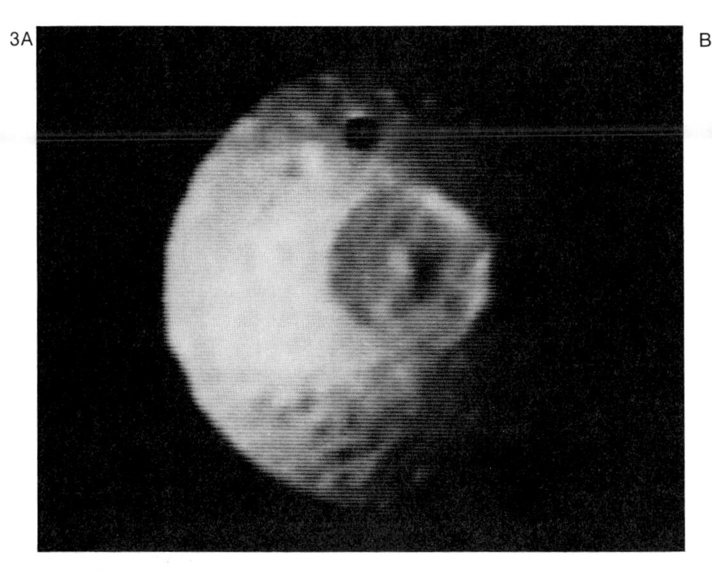

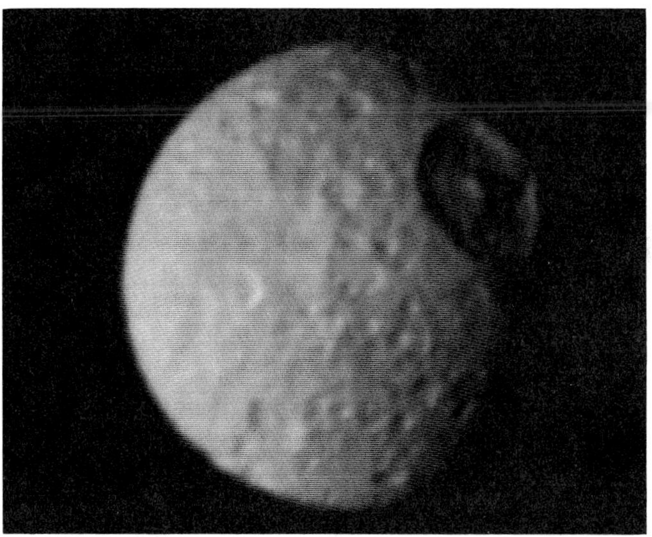

3. Mimas' giant crater
This well-formed structure—130 km in diameter—is almost centered on Mimas' leading face. The fact that this feature has a diameter almost one third that of the satellite itself, which can be appreciated from this Voyager 1 image (**A**) taken from a range of about 660,000 km, means that Mimas has one of the largest crater diameter/satellite diameter ratios in the Solar System. The second photograph (**B**), taken closer in at a range of 425,000 km, shows up dramatically the crater walls, which rise to a height of 5 km above the floor, the central peak, prominent in both images, rises to 6 km above the floor.

Map of Mimas

Name	Position
Arthur	190°W,35°S
Balin	82°W,22°N
Ban	149°W,47°N
Bedivere	145°W,10°N
Bors	165°W,45°N
Dynas	75°W,8°N
Elaine	102°W,44°N
Gaheris	287°W,46°S
Galahad	135°W,47°S
Gareth	280°W,44°S
Gwynevere	312°W,12°S
Herschel	104°W,0°
Igraine	225°W,40°S
Iseult	35°W,48°S
Kay	116°W,46°N
Launcelot	317°W,10°S
Lot	227°W,30°S
Mark	297°W,28°S
Merlin	215°W,38°S
Modred	213°W,5°N
Morgan	240°W,25°N
Pellinore	128°W,35°N
Percivale	171°W,1°S
Tristram	26°W,58°S
Uther	244°W,35°S

Name	Position
Avalon Chasma	120–160°W,20–57°N
Camelot Chasma	0–45°W,25–60°S
Oeta Chasma	105–130°W,10–35°N
Ossa Chasma	280–305°W,10–30°S
Pangea Chasma	290–340°W,25–55°S
Pelion Chasma	200–235°W,20–25°S
Tintagil Chasma	190–235°W,43–60°S

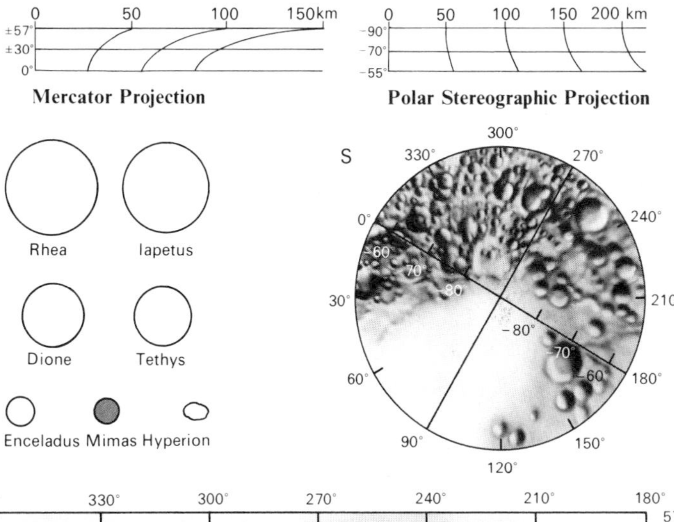

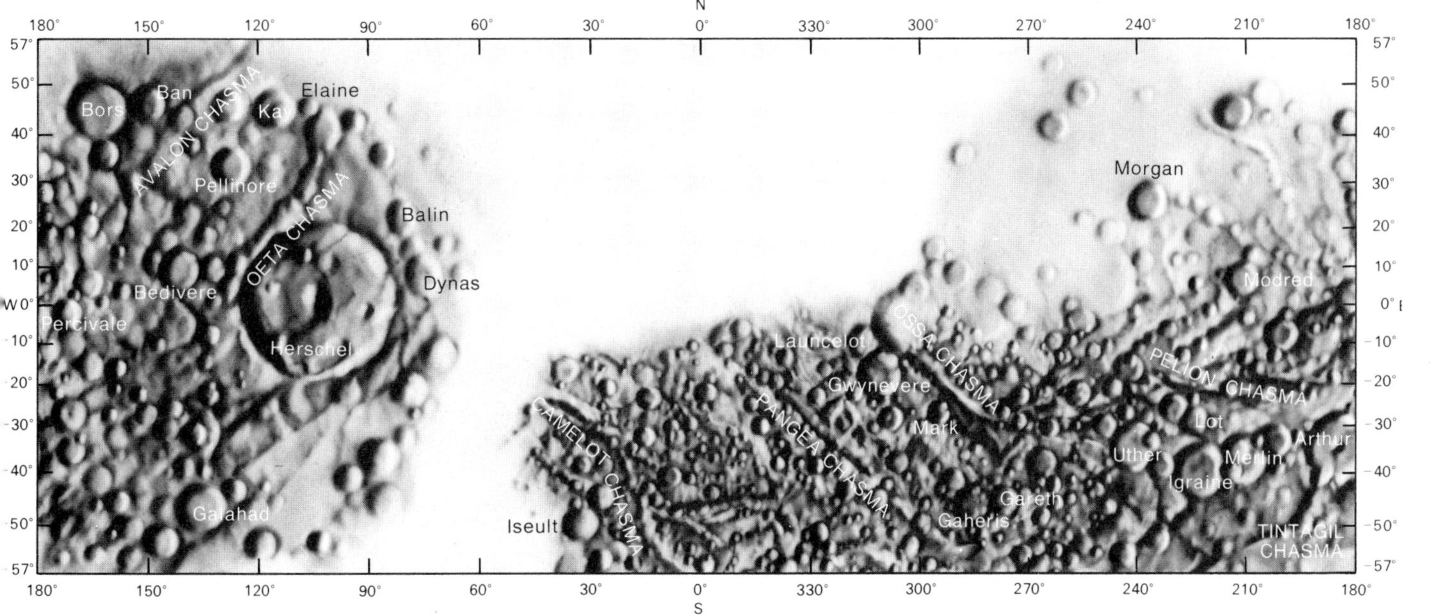

Enceladus

Enceladus is one of Saturn's smaller satellites: it shows the greatest geological evolution and is the most "youthful". Its diameter is only 500 km, less than half that of Tethys or Dione, and its density is low, only a little greater than that of water. The Voyager 1 images were obtained across a minimum distance of more than 200,000 km, but were detailed enough to show that the surface of Enceladus is different from those of the other icy satellites, for it is smooth. Voyager 2 passed Enceladus at 87,140 km. In general, Enceladus is bright; it is in fact the most reflective body in the entire Solar System, with an albedo not far short of 100 percent.

Six different types of terrain have been found. First there are the heavily cratered plains A and B, thought to be the oldest parts of the surface now on view. There are no giant features, but there are plenty of smaller craters in the 10–30 km diameter range. According to the impact theory of cratering, these formations were produced during the period when Enceladus was being bombarded by assorted debris orbiting Saturn. In region A many of the craters show evidence of collapse, and those with central peaks show very gentle, rounded mountains. In region B similar-sized craters are highly preserved, suggesting that the thermal histories of the two regions are different. The average depth of the craters in region B is greater. Regions C, D and E are of intermediate type, with linear features (valleys and ridges up to 1 km high) cutting through cratered plains. These make up the central part of the visible disc and the bowl-shaped craters are 5–10 km in diameter. But the most remarkable area of Enceladus is the extensive plain F, where there are virtually no craters at all, and the surface of the satellite is dominated by long grooves.

These variations in terrain suggest a complex geological history, during which the surface has been replaced in several stages. The crater plains on Enceladus have a crater density comparable to the least-cratered surfaces seen elsewhere on the Saturnian satellites, such as the smooth plains of Dione's leading hemisphere. The ridged plains, which are in the trailing hemisphere of Enceladus, have a crater density lower by a factor of 50. The present global mean cratering rate is estimated to be equal to the present lunar cratering rate by both asteroids and comets. The plains are thought to be 10^9 years old, which is only a quarter of the age of the satellite. Consequently a major resurfacing stage must have occurred on Enceladus relatively late in the geological history of this most unusual Saturnian satellite.

Valleys and ridges indicate crustal movements. They may have been formed by faulting accompanied by the extrusion of fluids. In the major region of the ridged plains, the ridges tend to form a concentric pattern near the border of the unit. The possible explanation is that they are pressure ridges formed by convective upwelling and the formation of new crust in the center of the region with compression and folding of the crust along the margins.

One possible cause of the exceptional surface of Enceladus is to be found in its connection with Dione, which is much larger, denser and more massive. Dione's orbital period is twice that of Enceladus, and the gravitational pull of Dione keeps the orbit of Enceladus slightly elliptical. This produces tidal effects, and it may well be that these are sufficient to keep the interior comparatively warm, so that material can be extruded from faults in the surface – presumably in the form of soft ice. Enceladus, unlike the other icy satellites, may be active, a surprising feature for so small a body. Such a situation is not unique, however; the same phenomenon is evident with Jupiter's inner Galilean satellite Io. It is unlikely that there is significant heating of Enceladus from radionuclide decay.

The E Ring of Saturn appears to exhibit a pronounced peak in brightness along the orbit of Enceladus and may consist chiefly of particles that have escaped from the satellite. If the surface of Enceladus were punctured, by meteoritic impact for example, water would outgas, forming supercooled droplets and ice crystals, which would escape from the weak gravitational field of Enceladus.

1

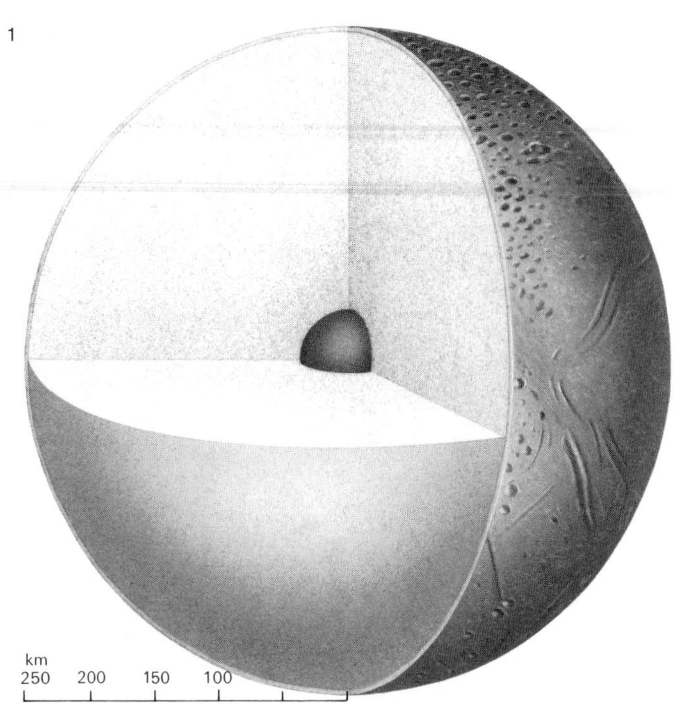

1. Interior of Enceladus
Like most of Saturn's satellites, Enceladus is believed to have a roughly 60:40 ice/rock ratio. It has, however, the most extensive geological history. One theory is that tidal forces exerted by Dione and Saturn produce heating in the satellite's interior. This may in turn result in outgassing of water, and perhaps methane as well, onto the ice-crusted surface.

2. Terminator
This view of the terminator of Enceladus shows up craters on the border of a cratered terrain and the ridged plain. At the lefthand edge of the cratered terrain (known as CT1) craters are older than those of the area to the right (CT2), where craters are deep and bowl shaped. These two cratered regions have had different thermal histories.

2

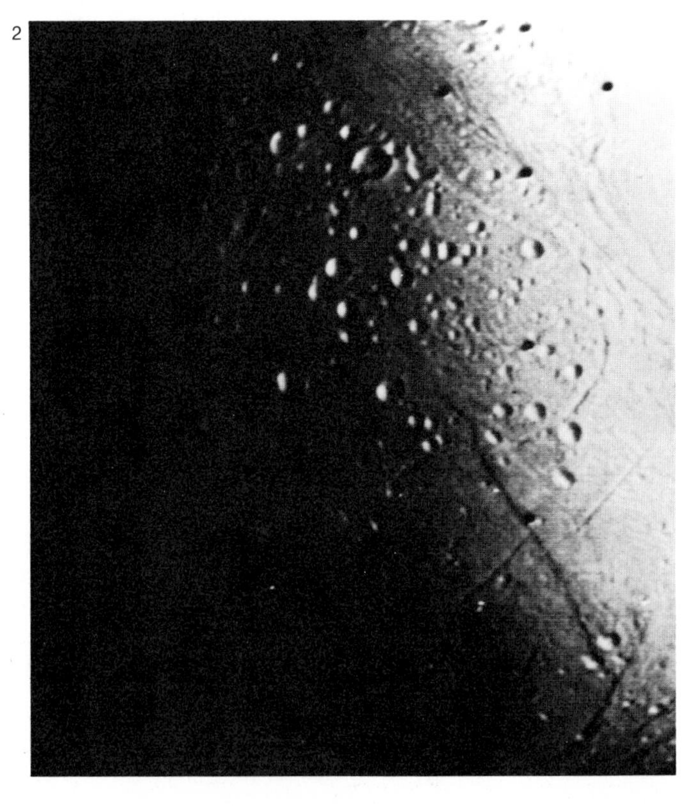

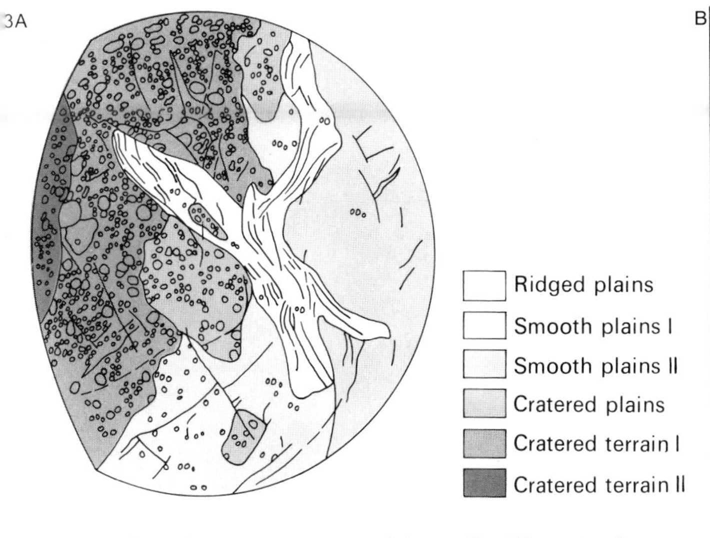

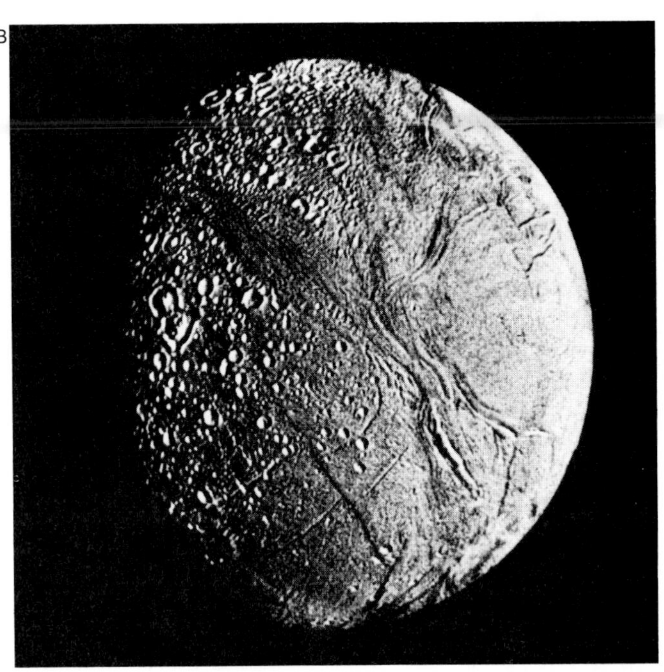

3. Types of terrain
The distribution of the six types of terrain found on Enceladus are shown on the diagram (**A**). Cratered terrains are the oldest, possibly from early bombardment of the satellite. The cratered plains are younger but have a range of ages. The ridged plain is the youngest terrain. The Voyager 2 image (**B**) has been enhanced to show up topography.

Map of Enceladus

Ali Baba	11°W,55°N
Dalilah	244°W,53°N
Dunyazad	200°W,43°N
Julnar	340°W,54°N
Shahrazad	200°W,49°N
Shahryar	222°W,58°N

Bassorah Fossa	23–345°W,40–50°N
Daryabar Fossa	20–335°W,5–10°N
Isbanir Fossa	0–350°W,10°S–20°N
Diyar Planitia	250°W,0°
Sarandib Planitia	300°W,5°N
Harran Sulci	210–270°W,5°S–35°N
Samarkand Sulci	300–340°W,10°S–75°N

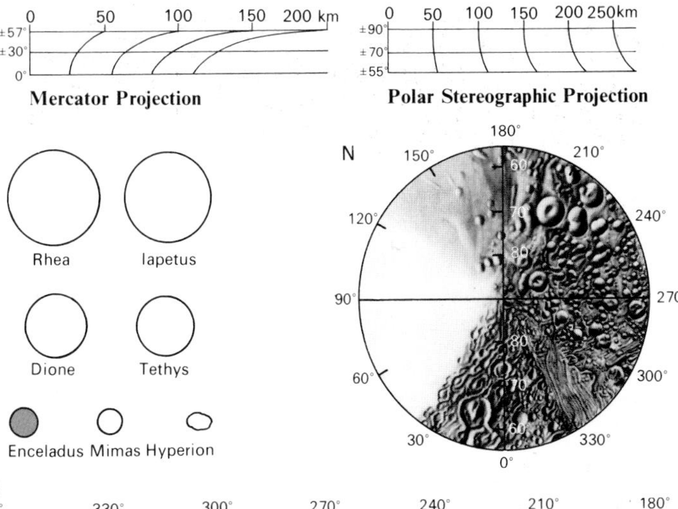

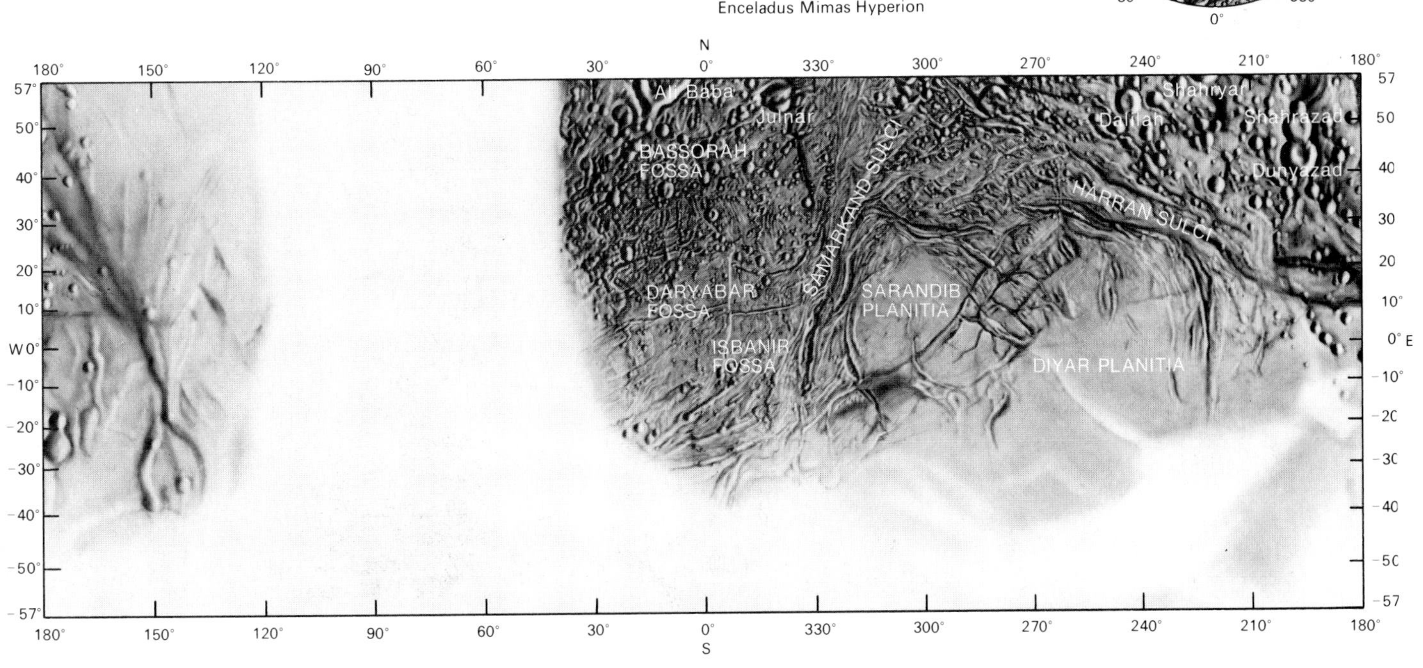

Map of Tethys

Ajax	285°W,30°S
Anticleia	38°W,55°N
Circe	49°W,8°S
Elpenor	268°W,54°N
Eumaeus	47°W,27°N
Eurycleia	247°W,56°N
Laertes	60°W,50°S
Mentor	39°W,3°N
Nestor	58°W,57°S
Odysseus	130°W,30°N
Penelope	252°W,10°S
Phemius	290°W,12°N
Polyphemus	285°W,5°S
Telemachus	338°W,56°N
Ithaca Chasma	30–340°W,60°S–50°N

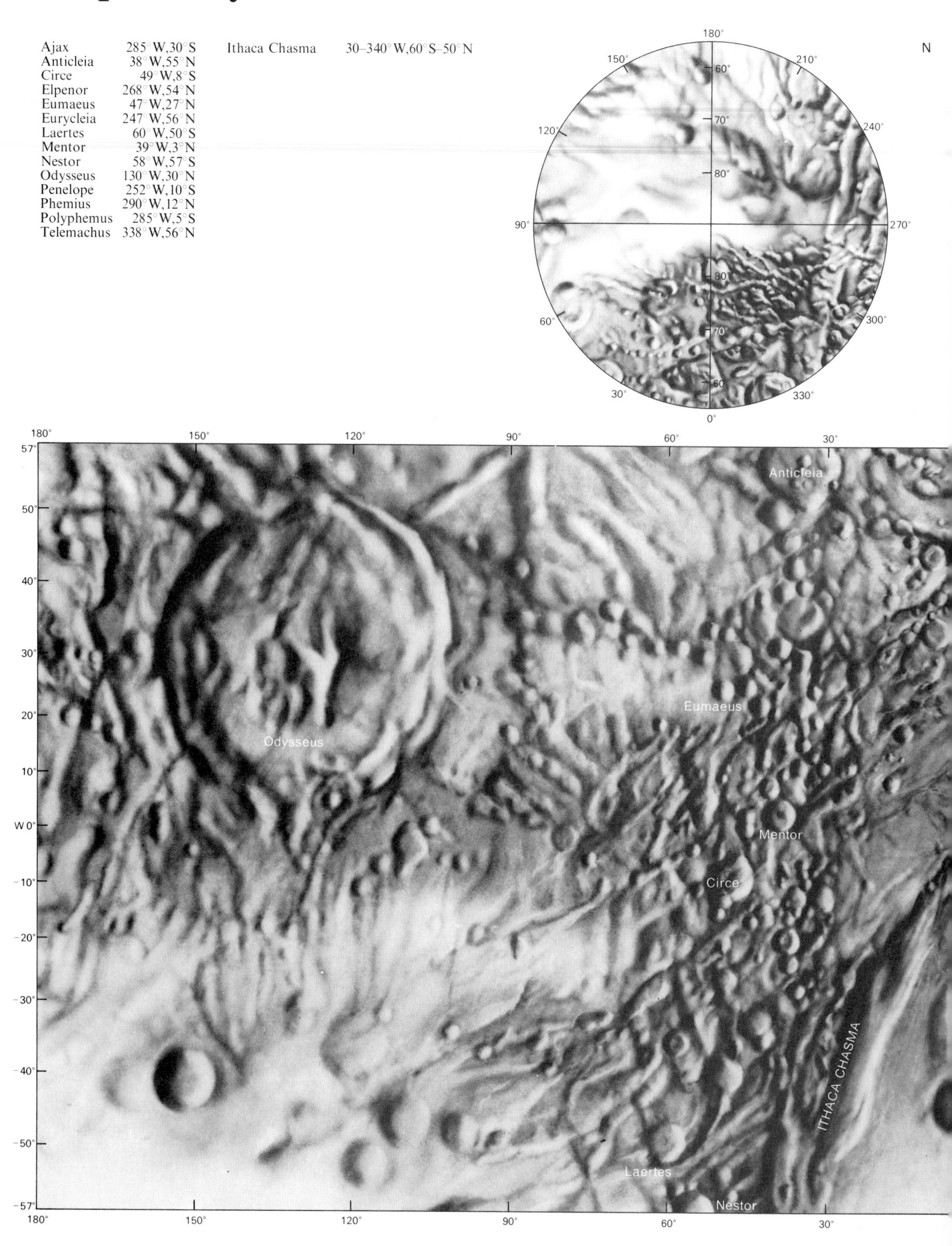

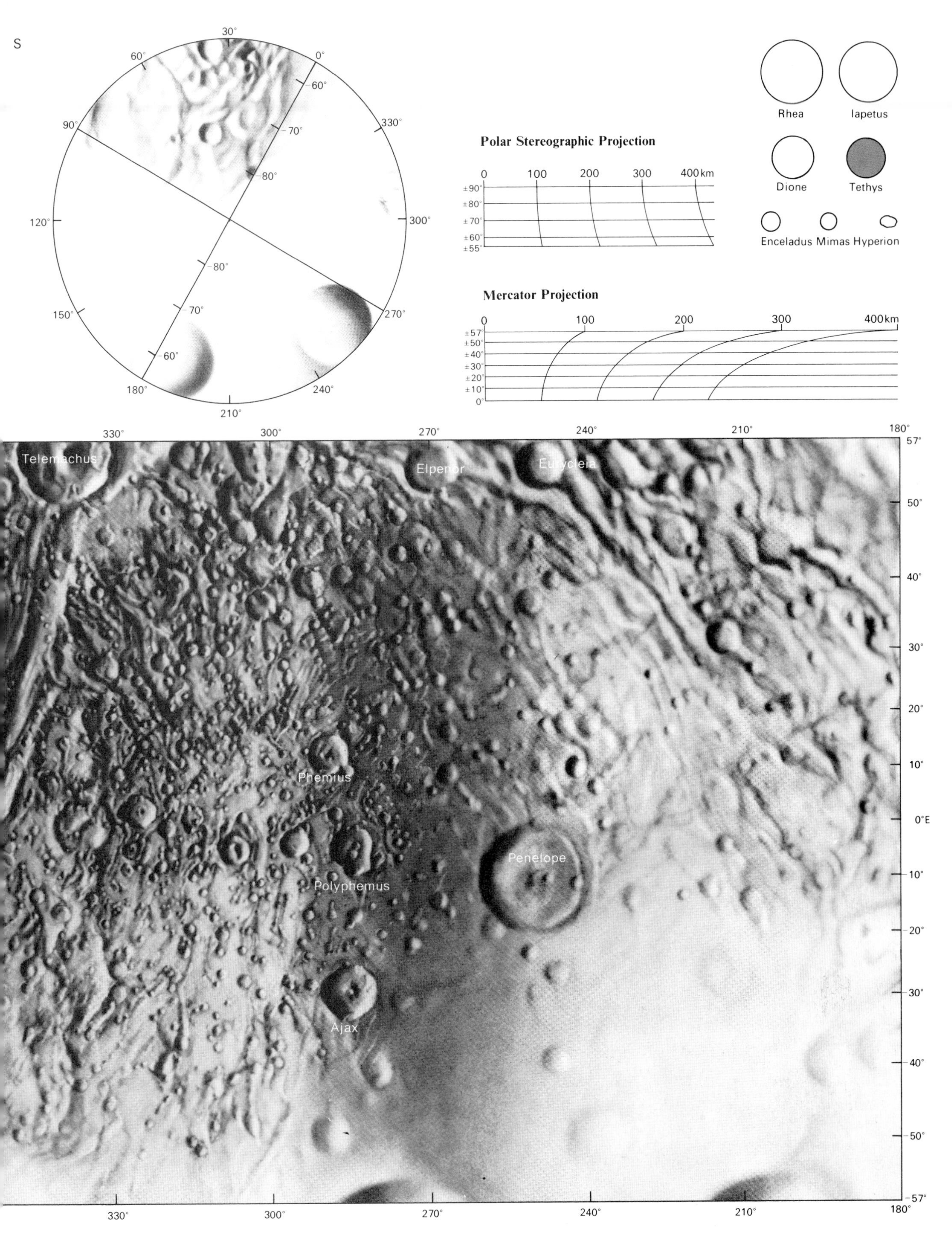
S
30°
60°
0°
-60°
90°
-70°
330°
-80°
120°
300°
-80°
-70°
150°
270°
-60°
180°
240°
210°
Polar Stereographic Projection
0 100 200 300 400 km
±90°
±80°
±70°
±60°
±55°
Mercator Projection
0 100 200 300 400 km
±57°
±50°
±40°
±30°
±20°
±10°
0°
Rhea
Iapetus
Dione
Tethys
Enceladus Mimas Hyperion
330° 300° 270° 240° 210° 180°
57°
50°
40°
30°
20°
10°
0°E
-10°
-20°
-30°
-40°
-50°
-57°
Telemachus
Elpenor
Eurycleia
Phemius
Polyphemus
Penelope
Ajax

Tethys

Tethys, the innermost of the larger satellites, has a diameter of 1,050 km. In size it is almost a twin of Dione, but is different in nature, mainly because of its exceptionally low density, which is about the same as that of water. Tethys seems to be made up of almost pure ice, and it has an albedo of 0.8, higher than that of any of the other satellites apart from Enceladus. Very little was revealed on Tethys by Voyager 1 apart from the huge Ithaca Chasma, but Voyager 2 bypassed the satellite at only 93,000 km, enabling much of the surface to be mapped. The Voyagers also discovered two small co-orbital satellites, one 60° ahead of Tethys and the other 60° behind. Both are small and very faint and would not have been detected from Earth.

There are two surface features of paramount importance. One of these is Ithaca Chasma, which is unlike anything else in the Solar System. It is a huge trench, extending around Tethys' globe from near the north pole down to the equator and down to the neighborhood of the south pole. Its average width is 100 km; it is 4–5 km deep, while the slightly raised rim reaches a height of approximately 0.5 km. Its origin is decidedly uncertain. It has been suggested that in its early stages of evolution Tethys consisted of a globe of liquid water with a thin solid crust; if so, the freezing of the interior would have caused expansion, producing the huge trough as a surface crack. Why only one crack was produced instead of a whole system of smaller ones has not been explained. Ithaca Chasma is definitely not a crater-chain of any kind, and it does indicate that there was a catastrophic event upon Tethys in the distant past.

Craters

The second feature of special interest is an enormous crater, about 400 km in diameter, lying at approximately latitude 30°N and longitude 130°. Its diameter is about 40 percent of that of Tethys itself, and it is larger than the whole of Mimas. It has a central peak, and is the largest crater with a well-developed central peak so far found in the Solar System. The crater floor may have rebounded tens of kilometers above the subdued rim. The surface of the interior follows the curve of the mean radius of Tethys, so that however it may have been formed it is a very ancient feature. It is much less well preserved than the large crater on Mimas, and there have been suggestions that beneath it there is a slightly warmer layer which has caused the crater to become flattened. It is not easy, however, to visualize such a layer in a body made up of pure ice, as Tethys appears to be. Whether it is in any way associated with the origin of Ithaca Chasma is a matter for debate, but that would appear to be rather doubtful.

There is another large crater to the north of the equator, at longitude 250°, with a central peak. Elsewhere there are very heavily cratered regions, with alignments that seem to be very approximately parallel to the overall direction of Ithaca Chasma. The region north of the equator is hilly and densely cratered.

Part of the trailing hemisphere exhibits a less rugged surface, with a few large craters surrounded by plains with a considerably lower crater density than the hilly terrain. The plains were probably produced by a flood of material that erupted from the interior. The flood may have surrounded large, preexisting craters, and overlapped their associated rim deposits. Tethys, then, appears to be an "ice ball" that bears the scars of activity on the surface that happened early in the satellite's history.

It has been tentatively suggested that in the past Tethys suffered such intense bombardment that it was actually broken up, after which its fragments came together again. In this case the whole satellite would be extremely fragile, but the theory is highly speculative, and rests upon no certain evidence. Undoubtedly the two principal features raise interesting questions, and in its way Tethys is as puzzling as some of the other members of Saturn's extensive family of satellites.

1

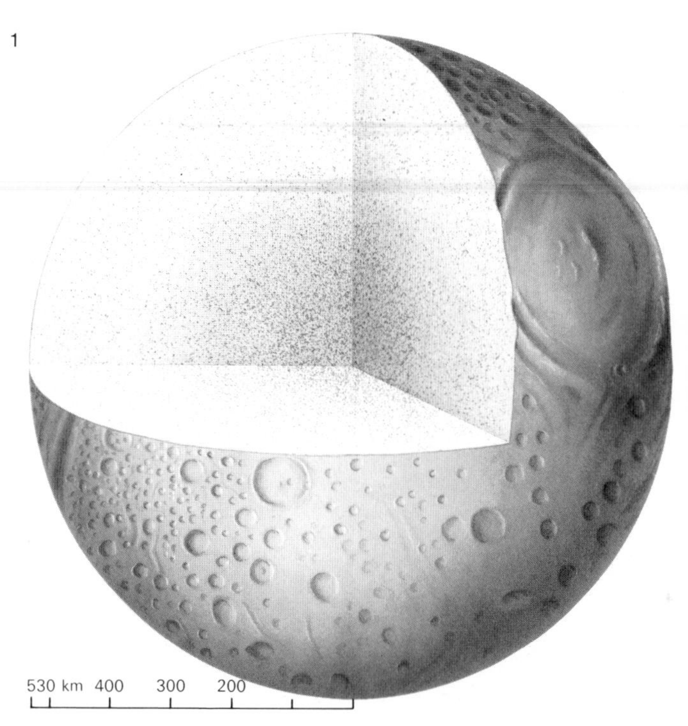

1. Interior of Tethys
Tethys is another low-density satellite thought to be almost entirely composed of water ice. There is evidence of resurfacing despite Tethys' small size. Any melting and surface alteration resulting from internal processes probably occurred after the main cratering. Less-cratered regions may well result from outpouring from the interior.

2. Ithaca Chasma
This enormous trench is centered on the Saturn-facing hemisphere of Tethys and extends for more than 270° around the body. It is not indicative of internal activity, but probably formed as the watery interior froze during Tethys' evolution. This would have created a 5 to 10 percent increase in surface area which matches the floor of the rift zone.

2

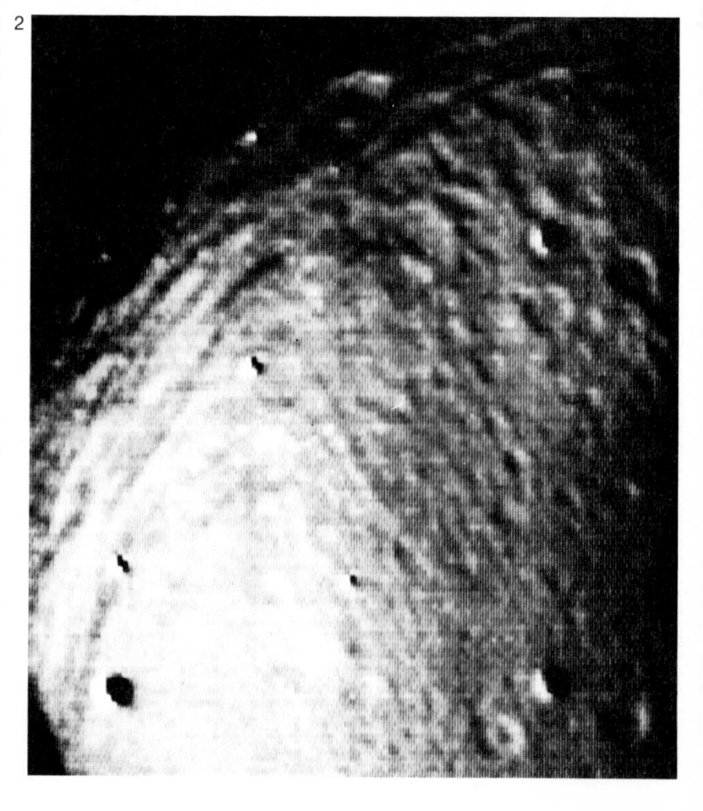

3. Giant crater
A series of photographs taken at 4-hourly intervals during Voyager 2's approach of Tethys shows the large, flattened crater—400 km in diameter—that dominates Tethys' other hemisphere. The second large crater can also be seen on the third image.

4. Global view of Tethys
The best global view of Tethys (Voyager 2) shows the trailing hemisphere and illustrates a variety of terrains.

5. Terminator
This was the highest-resolution image that was taken of Tethys before a spacecraft malfunction prevented further imaging of this satellite. It shows the hilly, heavily cratered region north of the equator and from this image a distribution of craters with diameters as little as 5 km has been determined on Tethys.

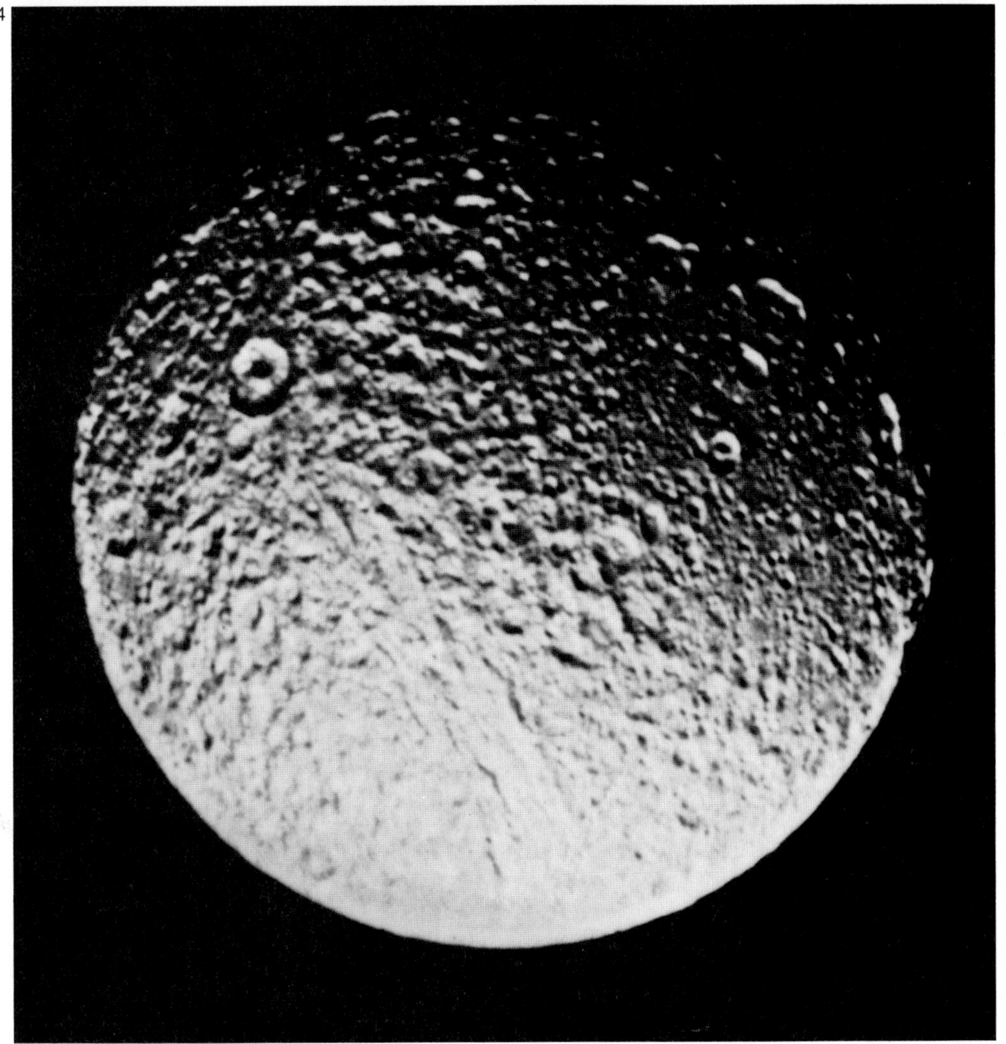

Map of Dione

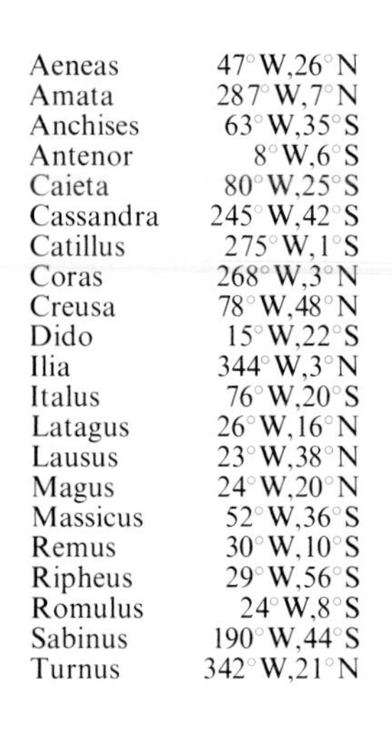

Aeneas	47°W,26°N
Amata	287°W,7°N
Anchises	63°W,35°S
Antenor	8°W,6°S
Caieta	80°W,25°S
Cassandra	245°W,42°S
Catillus	275°W,1°S
Coras	268°W,3°N
Creusa	78°W,48°N
Dido	15°W,22°S
Ilia	344°W,3°N
Italus	76°W,20°S
Latagus	26°W,16°N
Lausus	23°W,38°N
Magus	24°W,20°N
Massicus	52°W,36°S
Remus	30°W,10°S
Ripheus	29°W,56°S
Romulus	24°W,8°S
Sabinus	190°W,44°S
Turnus	342°W,21°N

Larissa Chasma	15–65°W,20–48°N
Latium Chasma	64–75°W,3–45°N
Tibur Chasmata	60–80°W,48–80°N
Carthage Linea	310–337°W,10–20°N
Padua Linea	190–245°W,5°N–40°S
Palatine Linea	285–320°W,10–55°S

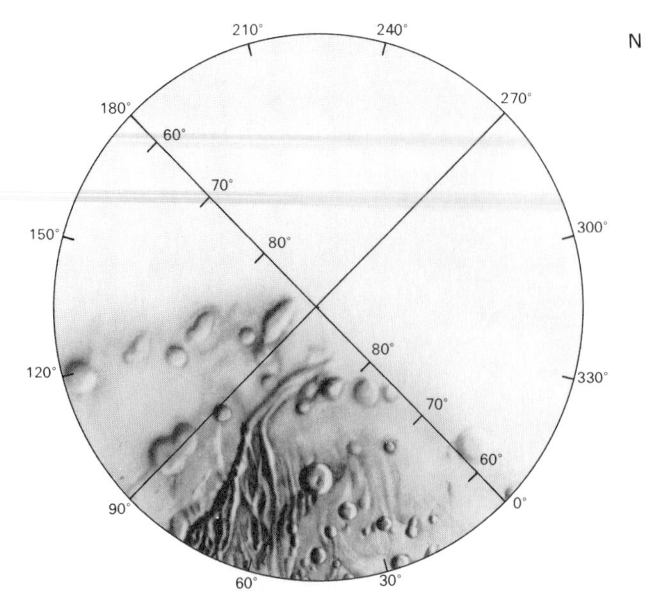

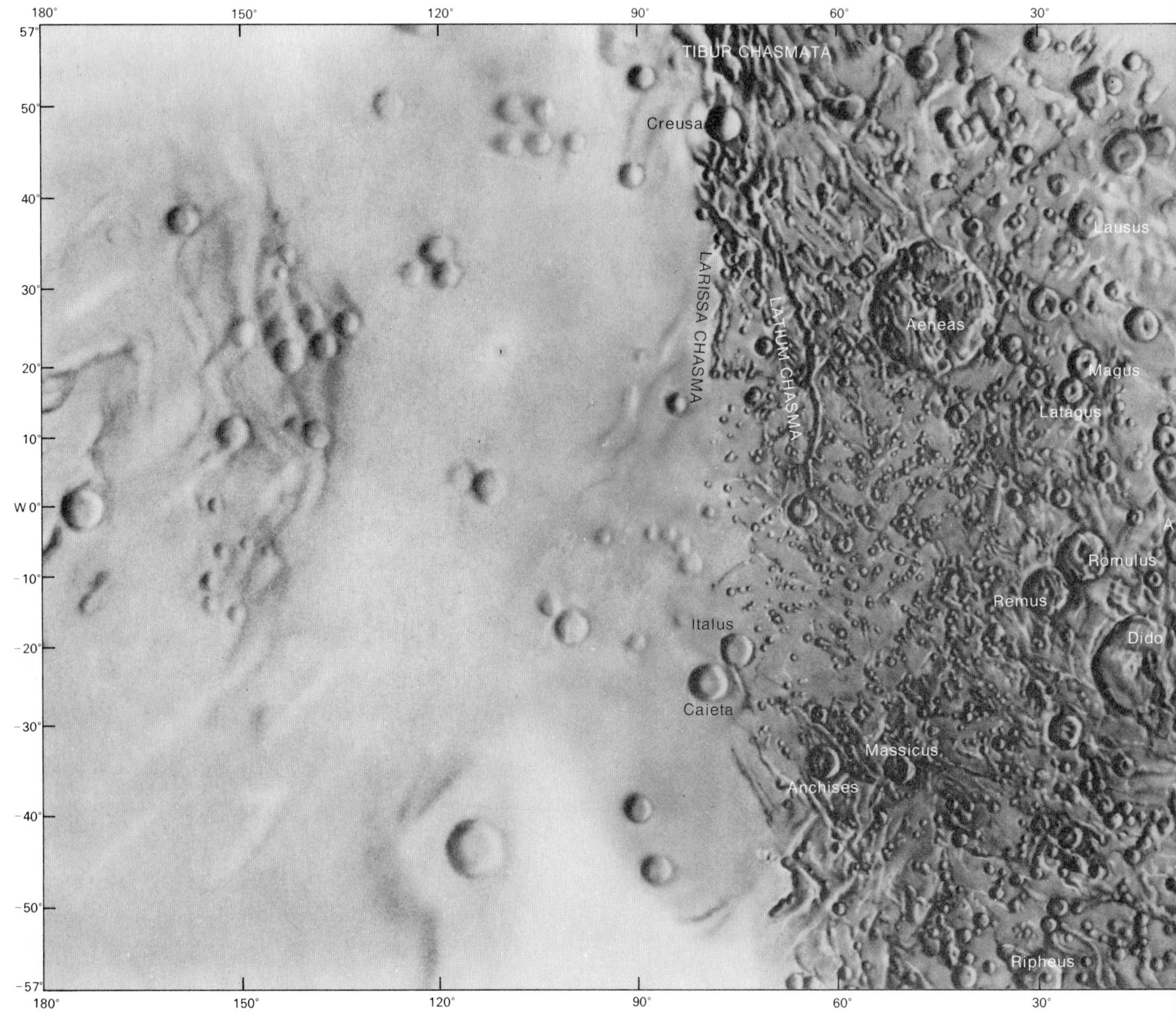

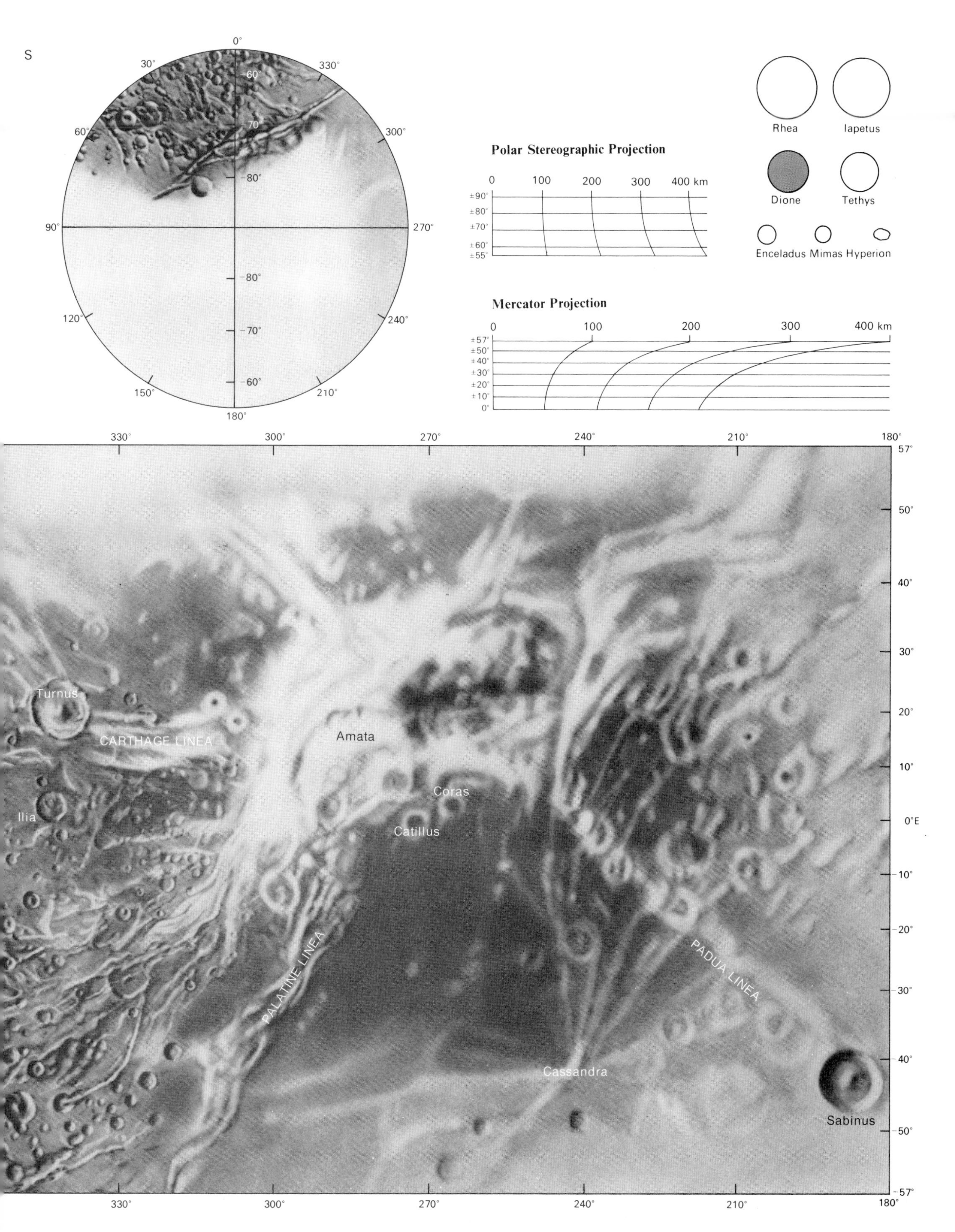
S
Polar Stereographic Projection
0 100 200 300 400 km
±90° ±80° ±70° ±60° ±55°
Mercator Projection
0 100 200 300 400 km
±57° ±50° ±40° ±30° ±20° ±10° 0°
Rhea
Iapetus
Dione
Tethys
Enceladus Mimas Hyperion
Turnus
CARTHAGE LINEA
Amata
Ilia
Coras
Catillus
PALATINE LINEA
PADUA LINEA
Cassandra
Sabinus
330° 300° 270° 240° 210° 180°
57° 50° 40° 30° 20° 10° 0°E -10° -20° -30° -40° -50° -57°

Dione

Dione is unusual in several respects. Its diameter, 1,120 km, is only slightly greater than that of Tethys, but its albedo is distinctly less, so that as seen from Earth it appears no brighter. Its density, 1.4 times that of water, is considerably greater than for any of the other named satellites apart possibly from Phoebe, and it has been suggested that Dione may affect both the radio emissions from Saturn and the surface of Enceladus, although the latter idea admittedly rests upon fragmentary evidence. The surface features of Dione also differ from those of the other satellites. It was surveyed both by Voyager 1 and, less effectively, by Voyager 2. A considerable portion of the total surface has been studied, although there is a gap between longitudes 80° and about 200°. Some features have been recorded in each polar region. Dione has one small co-orbital satellite, 1980 S6 (Dione B), which moves 60° ahead of it, and another has recently been reported after further studies of the Voyager results.

An important characteristic of Dione is that the brightness of the surface is far from uniform. The trailing hemisphere is comparatively dark, with an albedo of only about 0.3, while the brightest features on the leading hemisphere have an albedo of about 0.6. Only Iapetus, in Saturn's satellite system, displays greater differences in surface brightness.

The most striking feature on Dione is Amata, which lies at the center of a system of bright, wispy features that divide up the dark trailing hemisphere. They appear to be associated with narrow linear troughs and ridges that are extensions of bright lines. These features are possibly the result of internally generated stresses such as those that might have been produced by the freezing of the interior. the bright material, which is probably water ice, may have extruded from inside the satellite. Amata has a maximum diameter of about 240 km and its exact nature is uncertain: it may be an irregular crater, or it may be more in the nature of a basin; images so far have not been clear enough to tell. It is not the focus of a lunar-type ray system, indeed there are virtually no ray-craters on Dione. The wispy features are presumably faults or fractures, and there can be little doubt that Amata is related to their formation, but these features remain something of a mystery.

1

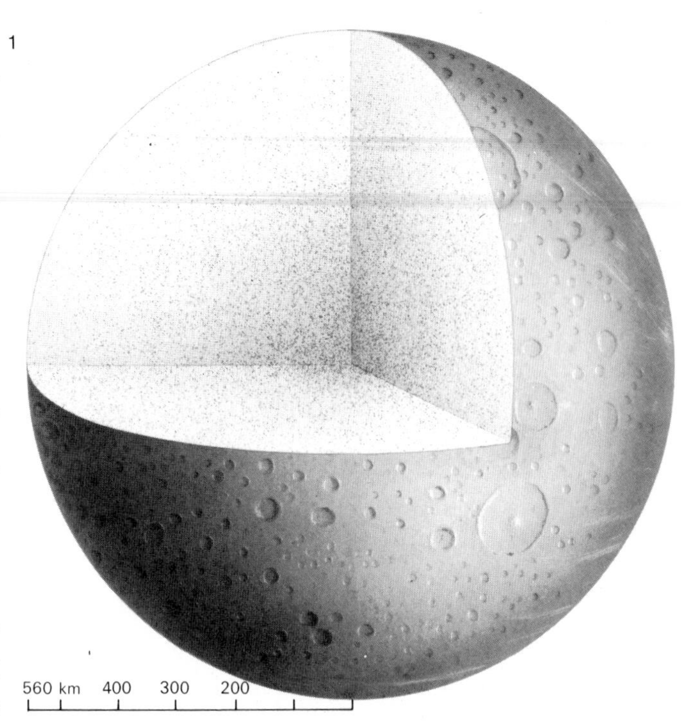

1. Interior of Dione
Dione, though roughly the same size as Tethys, is much denser and is believed to have a higher rock content. If the bright streaks represent an outpouring from the interior, then there was once, and possibly still is, an internal heat source. A relatively high concentration of radiogenic nuclides may be responsible and could also account for the satellite's higher density.

2. Dione's trailing hemisphere
From a range of 790,000 km, this Voyager 1 photograph shows the dramatic contrast between light and dark areas of Dione's trailing hemisphere. The bright, wispy features are thought to be surface frost. The bright bands, which are brighter than the brightest features on Jupiter's satellites, may be the result of internal activity at some time during Dione's geological history.

Craters

There are not many craters on Dione with diameters of more than 30 km, though a few have 40 km diameters. The larger craters—up to 165 km in diameter—seem to be decidedly shallower than those on Tethys, but some of them have pronounced central peaks. Craters are most numerous on the plain that covers much of the observed part of the region between longitudes 0° and 70°, though the Voyager coverage has not revealed whether or not the crater frequency beyond longitude 70° is equally great. It is thought that this is the most ancient part of Dione's surface: the brighter regions have presumably been resurfaced with a layer of material thick enough to conceal any craters that had been formed there. This may indicate that a former internal heat source kept Dione at least moderately active until a relatively late stage in the satellite's evolution. A high concentration of material undergoing radioactive disintegration may be the heat source. This would be consistent with Dione's density. The satellite is considered by some to be intermediate in type between Enceladus on the one hand, and Mimas and Tethys on the other, although the appreciably greater density of Dione must not be forgotten.

In the southern part of the heavily cratered plain there are a number of broad, low ridges trending in a northeasterly direction, and there are well-marked valleys. Close to the south pole there is a long linear valley, which has a length of about 500 km. The whole of the observed south polar region is thickly cratered; some of these craters have central peaks. In conclusion, Dione shows a very different aspect according to the hemisphere which is being presented to the observer.

2

3

4

3. Dione's bright features
The Saturn-facing hemisphere of Dione is seen here at a range of about 240,000 km (from Voyager 1). The bright radiating features may be the rays of debris thrown out from visible craters. Other bright areas might be ridges and valleys. The irregular valleys are old fault troughs that have been degraded by impacts.

4. Craters and faults
The large crater is less than 100 km in diameter and has a well-developed central peak. The valley is probably a fault line.

6

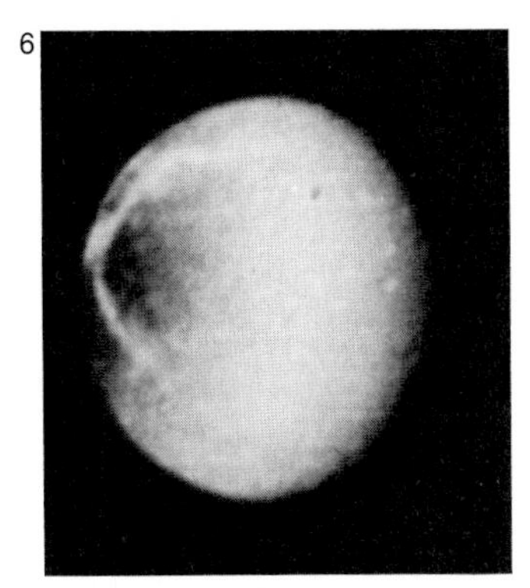

5

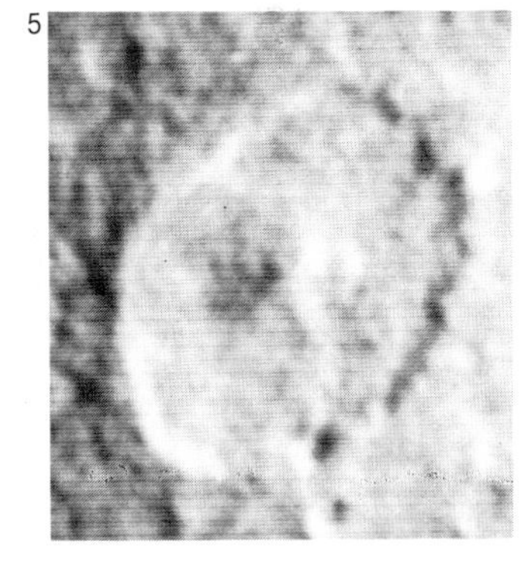

5. Crater close-up
This crater—less than 100 km in diameter—was photographed by Voyager 1 from a range of 162,000 km. This feature is part of an area of Dione showing many similar impact craters.

6. Dione and Saturn
This Voyager 1 photograph, taken from a range of 377,000 km, shows the satellite against a backdrop of Saturn itself. This view of the anti-Saturn face of Dione shows up well the darker trailing hemisphere and the brighter leading hemisphere.

Map of Rhea

Name	Position	Name	Position
Aananin	330°W,39°N	Pedn	340°W,48°N
Adjua	126°W,46°N	Qat	347°W,23°S
Arunaka	21°W,14°S	Sholmo	340°W,13°N
Atum	0°,45°S	Taaroa	99°W,14°N
Bulagat	14°W,35°S	Thunupa	15°W,51°N
Con	10°W,24°S	Tika	87°W,25°N
Djuli	46°W,26°S	Wuraka	357°W,28°N
Faro	121°W,52°N	Xamba	347°W,4°N
Haik	27°W,34°S	Yu-ti	85°W,55°N
Haoso	8°W,9°N		
Heller	310°W,9°N	Kun Lun Chasma	275–300°W,37–50°N
Iraca	120°W,45°N	Pu Chou Chasma	85–115°W,10–35°N
Izanagi	298°W,49°S		
Izanami	310°W,46°S		
Jumo	65°W,56°N		
Karora	16°W,7°N		
Khado	349°W,45°N		
Kiho	354°W,10°S		
Kumpara	321°W,11°N		
Leza	304°W,19°S		
Lowa	9°W,45°N		
Manoid	2°W,33°N		
Melo	6°W,51°S		
Num	93°W,23°N		

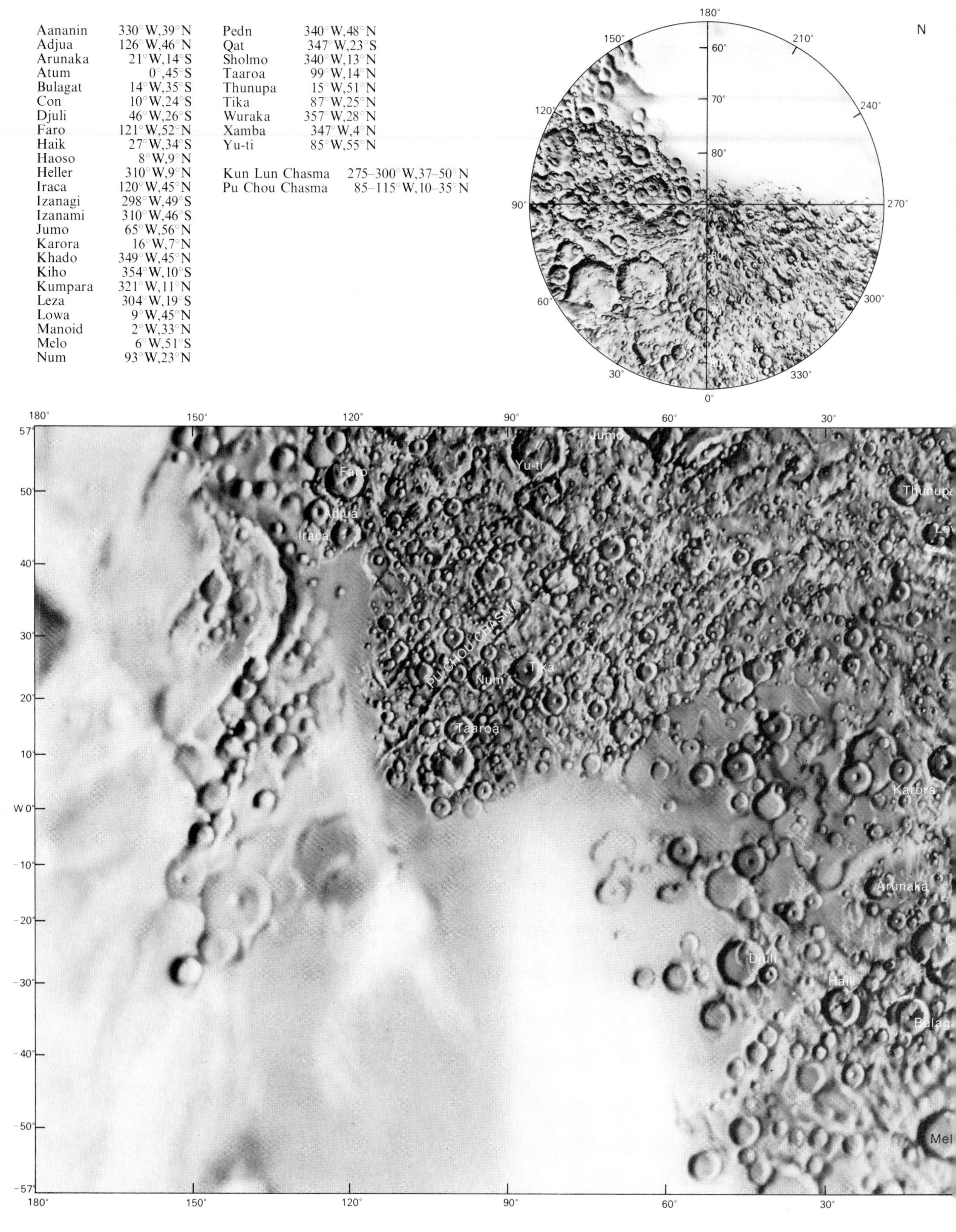

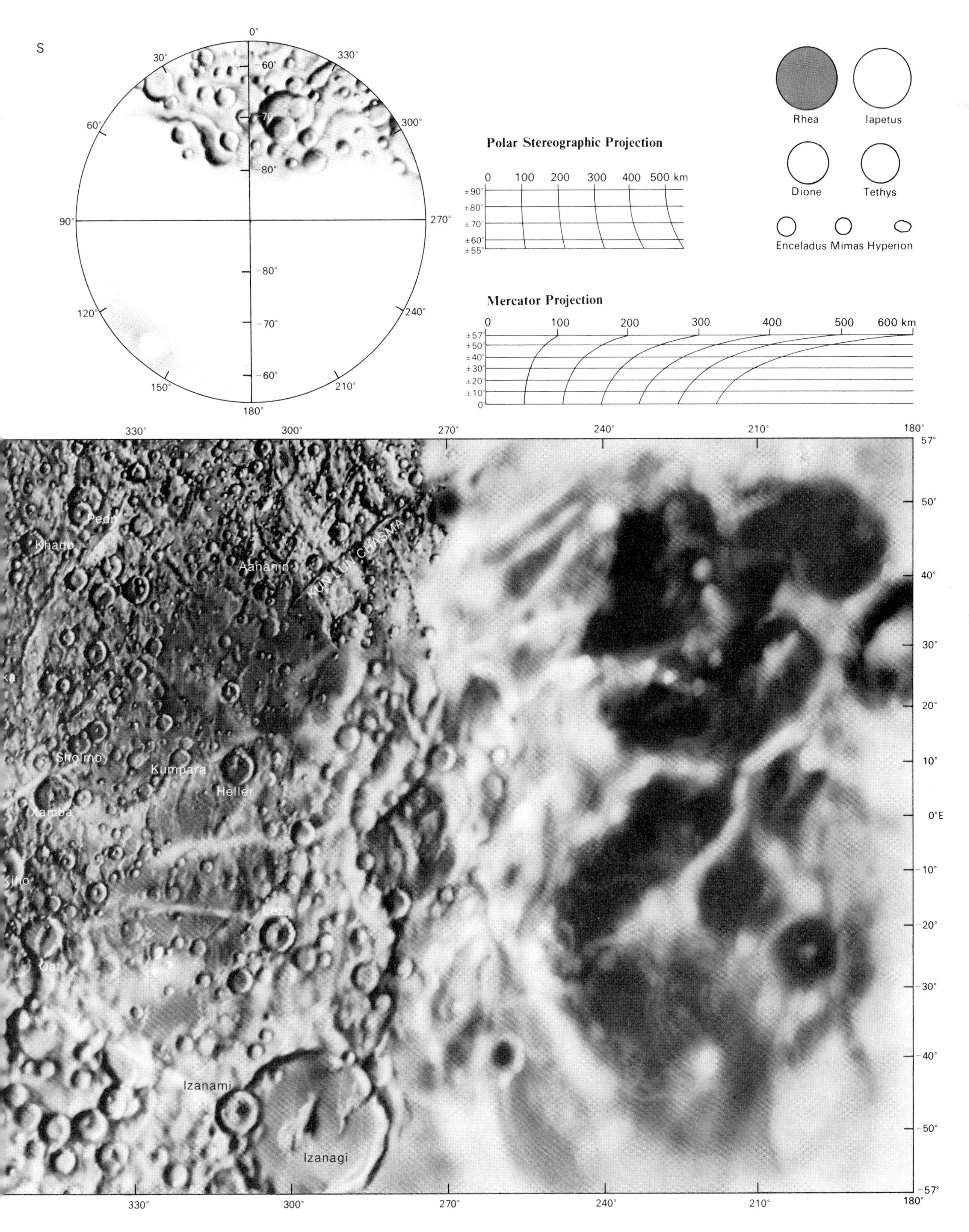
S
Polar Stereographic Projection
Mercator Projection
Rhea
Iapetus
Dione
Tethys
Enceladus Mimas Hyperion
KUN LUN CHASMA
Pedn
Khado
Aananin
Sholmo
Kumpara
Heller
Xamba
Leza
Qat
Izanami
Izanagi

Rhea

Rhea is the largest member of Saturn's system of satellites apart from Titan. Its diameter is 1,530 km, slightly greater than that of Iapetus, not slightly less, as was formerly believed. Its density, 1.3 times that of water, is somewhat less than that of Dione but considerably greater than that of Tethys and Iapetus. Voyager 1 bypassed Rhea at a distance of only 73,980 km, and obtained images of greater resolution than for any of the other icy satellites. Voyager 2 made its pass at more than 640,000 km, so it added little to Voyager 1's findings. A large part of Rhea's surface has been surveyed, including more than half the north polar region.

As with Dione and, even more markedly, with Iapetus, there is a pronounced difference between the two hemispheres of the satellite. The leading hemisphere is comparatively uniform and bland, though there is one diffuse feature near 90° which may possibly be a ray-center. This feature is unfortunately situated in an area of Rhea that was not studied at close range, so its precise nature remains uncertain. The trailing hemisphere is darker, and includes some wispy features that are not unlike those of Dione, though they are much less conspicuous; presumably they are the result of the same process, the exact nature of which is as yet uncertain. There is no doubt that the surface of Rhea is icy.

Craters

The equatorial region, particularly to the north of the equator, is densely cratered, and resembles the rolling cratered highlands of the Earth's moon. It has been commented that if these craters are the result of meteoritic impact, then Rhea may have suffered a terrible battering in the past. Yet the craters are not identical in type to those of some of the other satellites. Many of them are irregular, with a marked tendency to be polygonal in shape. There are no flattened craters. Rhea's low gravity and relatively small diameter permitted rapid cooling and freezing, thus preserving crater forms. The overall impression is that Rhea's outer crust is covered with a layer of what can only be described as rubble, and parts of the surface are extremely ancient even by Solar System standards. Craters follow the usual pattern of distribution, with small craters intruding into larger ones rather than vice versa, and some of the craters have prominent central peaks. Strings and chains of relatively small craters are common. There are few really large craters on Rhea.

Bright and dark areas

Two areas in the north polar region were better surveyed than any other part of Rhea's surface in as much as Voyager 1 passed right over them at the time of its closest approach. One bright and one dark area whose albedos differ by about 20 percent were found: they are reasonably well defined and the boundary between them lies approximately at longitude 315°. The bright terrain includes a greater number of comparatively large craters, some with fairly bright floors that have relatively little detail. The sharp boundary between the two areas suggests that the difference in crater density reflects differences in the ages of the areas.

According to the impact theory, there were two separate periods of intense bombardment. Between these two there was a period of resurfacing, and the second bombardment produced fewer large craters. In Rhea's case the dark material was extruded from below the surface, covering up the craters that had been produced by the initial bombardment. Localized bright patches seen inside some of the craters could be due to the exposure of comparatively fresh ice on the satellite's surface.

The equatorial region of the satellite includes several areas of low crater density near the equator (60–70° and 300–330°). There is a complex region of narrow linear grooves and troughs in the region from 290° to 330° W and 30° to 40° S. It ends abruptly on the western margin of an area of large craters. Rhea certainly exhibits evidence of a varied geological history.

1

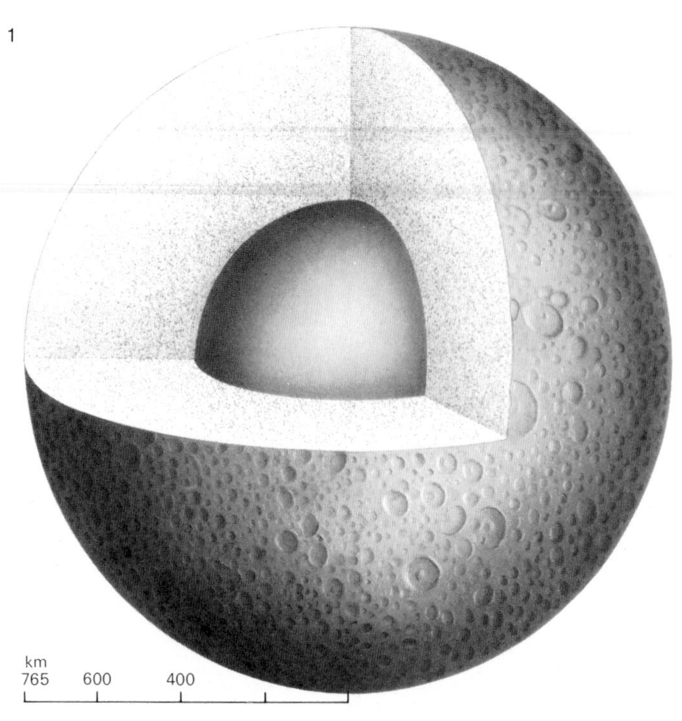

1. Interior of Rhea
According to some theories, the larger, more dense outer satellites of Saturn, such as Rhea, may have a core. It is believed that Rhea has a roughly 50:50 ice/rock composition, and at some stage in its thermal history rock constituents were differentiated. The core's size is a guesstimate.

2. Craters
From the relatively close range of 73,000 km, multiple craters were seen on Rhea by Voyager 1. Many are old and degraded by later impacts or crustal disturbances. Central peaks may have been formed by rebound of the crater floor. Crater diameters are up to 75 km.

2

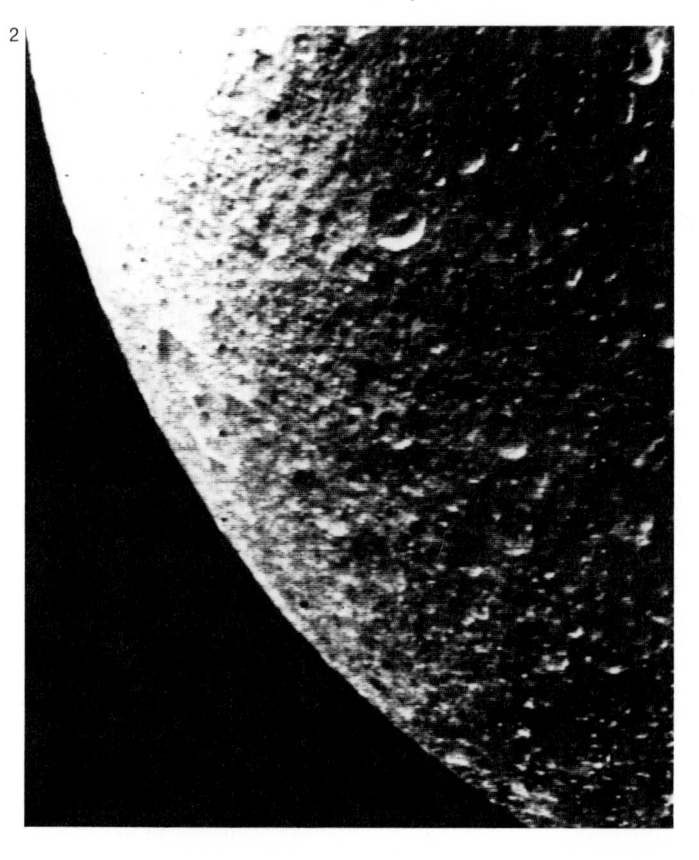

3A

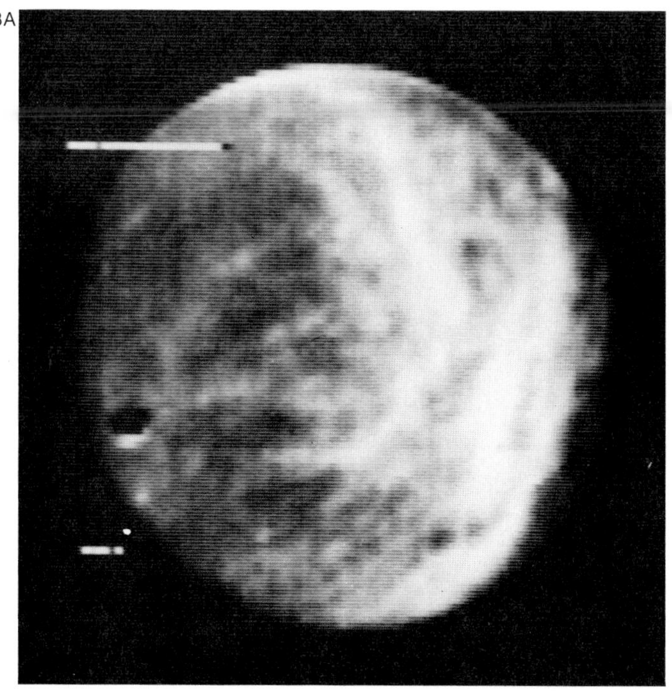

B

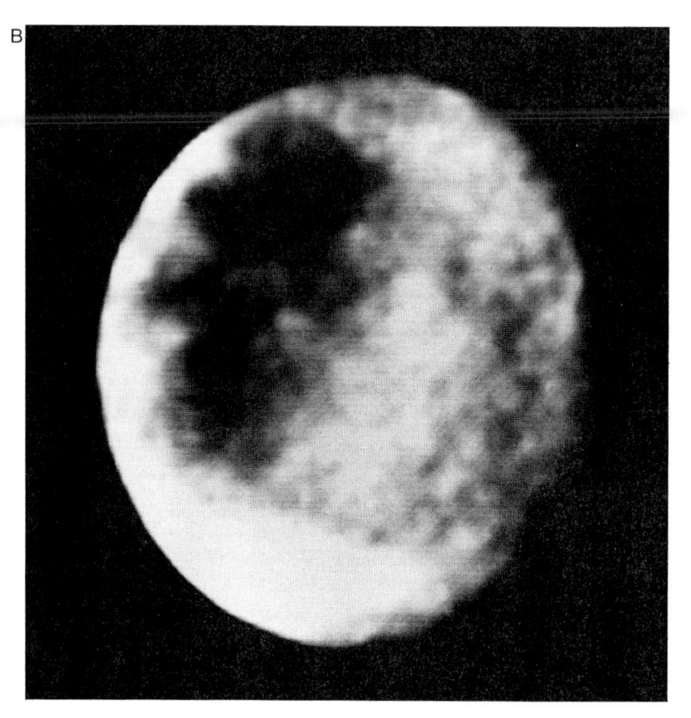

3. Bright and dark
Both these Voyager 1 images—taken from 1,925,000 km (**A**) and 2,700,000 km (**B**)—show broad bright areas against a darker background. The entire surface is thought to be covered by water frost and ice: the bright streaks may be pulverized ice particles thrown out from craters.

4. North polar region
A mosaic of Voyager 1 images at a distance of about 80,000 km shows the most heavily cratered of Saturn's moons. The largest in this picture is about 300 km in diameter. The lefthand diagram sketches light and dark areas and the right shows new craters (black) and mantled ones.

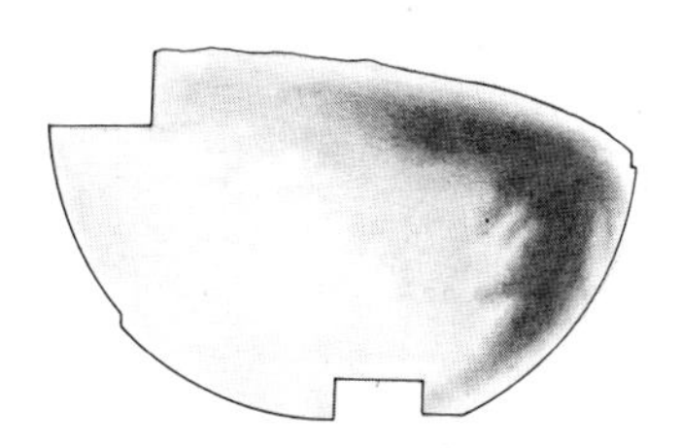

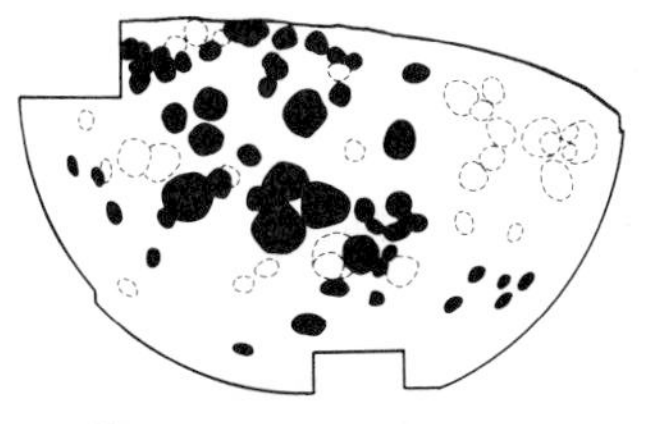

4

Titan I

Titan, by far the largest of Saturn's satellites, was formerly believed to be the largest satellite in the Solar System. Voyager observations have shown, however, that it is slightly inferior to Ganymede in Jupiter's system. The diameter of Titan is 5,150 km. The mean density of the body is about twice that of water, so that it may be assumed that it is probably composed of equal amounts by mass of rock and ice.

Titan's escape velocity is almost 2.5 km/sec. It is feasible, therefore, that at the distance of the Saturnian system from the Sun, Titan might retain an atmosphere. The first indications of this were given in 1903 by the Spanish astronomer J. Comas Solá, who found that there was considerable limb darkening. Observations of this kind are not easy, and cannot be conclusive, but in the winter of 1943–44 Gerard P. Kuiper showed spectroscopically that Titan has an atmosphere containing methane.

Constituents of the atmosphere

To the surprise of many investigators, Voyager 1 analyses showed that Titan's atmosphere is made up almost entirely of nitrogen, with less than one percent of methane in the upper part of it. It was found that no surface details were visible at all on the satellite, and the atmosphere was much denser than had been generally believed before the Voyagers. The surface pressure is 1.6 times that on the surface of the Earth, but conditions are very dissimilar; the temperature on Titan's surface is a mere 92 K. On the other hand, Titan and the Earth are the only bodies in the Solar System to have nitrogen-rich atmospheres.

In addition, Voyager 1 found that there is what may be termed "smog", or aerosols—products of the methane/ammonia chemistry. Acetylene, ethylene, hydrogen, methylacetylene and propane exist there (*see* diagram 1). Carbon dioxide has also been detected and carbon monoxide is suspected but has not yet been confirmed. The hydrocarbon concentrations on Titan are considerably higher than in the case of Jupiter and Saturn. The Titan/Saturn ratio for ethane is about 4, and for acetylene it is about 150; up to the present time ethylene has not been detected on Jupiter or Saturn.

On Titan, about 95 percent of the dissociated methane is irreversibly converted into acetylene and ethane. The constant removal of methane from the upper atmosphere by chemical reactions means that it must be constantly replenished from the surface. The temperature reaches a minimum at a level where the atmospheric pressure is about 200 millibars, and this acts as a "cold trap" for the methane and any other atmospheric constituent whose source is lower down. In effect, the tropopause regulates and maintains a constant supply of methane in Titan's stratosphere.

The atoms and molecules of hydrogen that are produced by photochemical reactions can easily escape from Titan because of the comparatively weak gravitational pull. However, they cannot escape from the pull of Saturn itself, and the result is a doughnut-like torus of hydrogen that remains in the orbit of Titan. Hydrogen is indeed widely distributed throughout the Saturnian system, and appears to form a cloud of uniform density encircling the planet between 8 and 25 R_s near the equatorial plane. Most of the hydrogen seems to be concentrated within 6 R_s. Whether or not this supply of hydrogen comes from Titan alone is not certain.

Although molecules of hydrogen cyanide have been observed in interstellar space, this is the first detection of them in a planetary atmosphere. This discovery is of major importance in as much as hydrogen cyanide is a key intermediate between the synthesis of amino acids and the bases present in nucleic acids. Inevitably this has led to the suggestion that Titan may support life, but this would be very improbable in such low temperatures as exist on Titan. On the other hand, it is true to say that Titan may be a sort of "Earth in deep freeze", and that the chemistry leading to the formation of organic molecules has occurred there even if the intense cold has prevented it from developing further.

Composition of Titan's atmosphere

	% volume
Hydrogen	0.94
Helium	0.06
Methane	1×10^{-2}
Ethane	2×10^{-5}
Acetylene	3×10^{-6}
Propane	2×10^{-5}
Diacetylene	10^{-8}–10^{-7}
Methylacetylene	3×10^{-8}
Hydrogen cyanide	2×10^{-7}
Cyanoacetylene	10^{-8}–10^{-7}
Cyanogen	10^{-8}–10^{-7}
Carbon dioxide	10^{-10}
Carbon monoxide	?

1

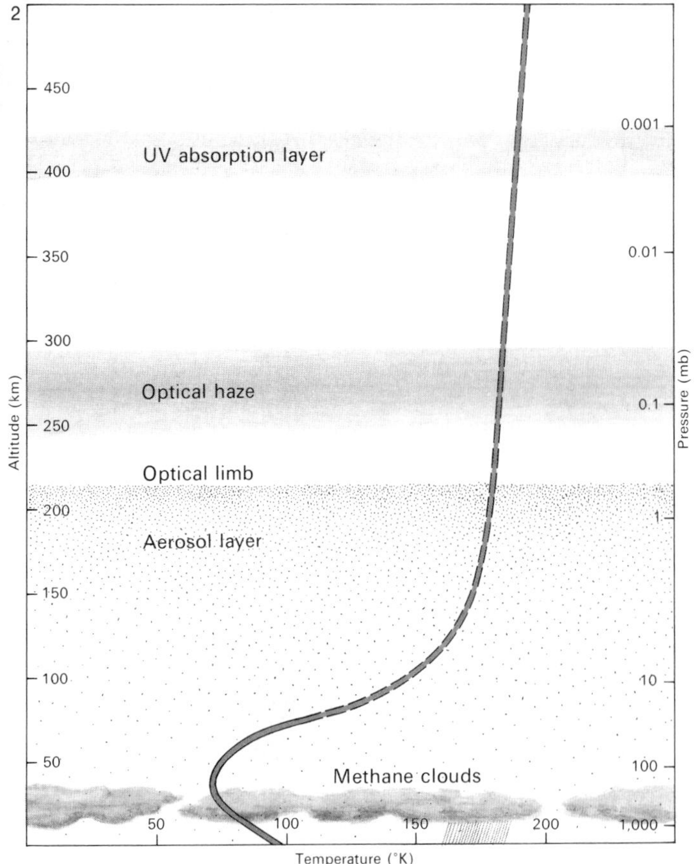

1. Titan's atmosphere
Visually, Titan was something of a disappointment, as no surface details were observed by either Voyager. This view of the night side of Titan from a range of 907,000 km shows the satellite's extended atmosphere, revealed by the scattering of sunlight.

2. Structure of the atmosphere
Voyager 1 dramatically increased our knowledge of Titan's dense atmosphere. Two previously unknown layers were discovered: an ultraviolet absorption layer that is transparent to light, and below this a haze layer. Beneath the haze is a layer of aerosol particles. Suspended at an altitude of some 200 km, the particles scatter sunlight in the atmosphere and obscure any small-scale contrast of light and dark features that might have been seen on Titan's surface. During the Voyager encounters no breaks were observed in the cloud cover. The aerosol particles are thought to aggregate, forming larger particles, which fall to the surface. Just above the surface are methane clouds and possibly methane rain, although as yet this has not been confirmed. During the Voyager 1 encounter the spacecraft was occulted by the atmosphere and readings obtained indicate a surface temperature of about 95 K and a pressure of 1,500 mb. A thermal inversion, as indicated in the diagram by the thick black line, is anticipated at high altitudes in the atmosphere.

The color of Titan

Titan's disc appears reddish or orange in color, but the two hemispheres are not identical. The southern hemisphere is relatively uniform in brightness, while the northern hemisphere is darker and redder. At the time of the Voyager 1 pass the north pole was covered by what appeared to be a dark hood; at the time of Voyager 2, nine months later, it looked more like a dark collar. There are certainly substantial differences in the atmospheric composition at the north pole and the mid-latitude regions; at lower latitudes there is considerably more acetylene and methylacetylene than in the polar zones.

The boundary between the two hemispheres lies in Titan's orbital plane, and this symmetry is almost certainly produced by the satellite's rotation, which is synchronous. The inclination of Titan's axis to the perpendicular to the orbital plane is believed to be 5° or less. Contrast enhancement of images shows up zonal features that resemble the belts and zones of Jupiter and Saturn.

These differences between the two hemispheres—detected by Pioneer 11 as well as both the Voyagers—are important in considering atmospheric processes. Changes in the chemistry of the upper atmosphere produced by the periodic entry into and exit from Saturn's magnetosphere (*see* page 20) do not seem to be the cause. The differences are probably due to seasonal effects, bearing in mind that the seasons on Titan are extremely long—about 7.5 years. First one hemisphere, then the other receives the greater amount of radiation from the Sun, and it is reasonable to assume that during Titanian summer the atmosphere is affected to a lower level. Some confirmation has been drawn from the slight but perceptible variations in the apparent brilliancy of Titan as seen from the Earth, which have been closely studied since 1972. Between 1972 and 1976 the brightness increased by 9 percent at blue wavelengths and about 5.5 percent at yellow, reaching a maximum during 1976–77 and then decreasing again during 1978.

3

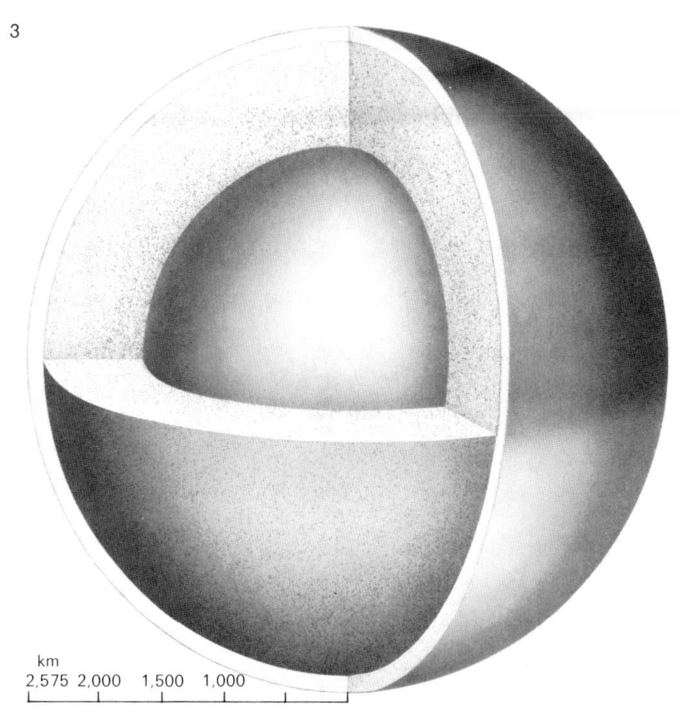

3. Interior of Titan
Titan has a density nearly twice that of water, which suggests that it is composed roughly of half rock and half water ice. It has a solid surface and the absence of an intrinsic magnetic field means that there is no liquid electrically conducting core. One model suggests that the structure of Titan is rather like that of Ganymede in Jupiter's system. This would mean that there was a soft ice mantle beneath the crust, and a relatively large, silicate core beneath that.

4

5

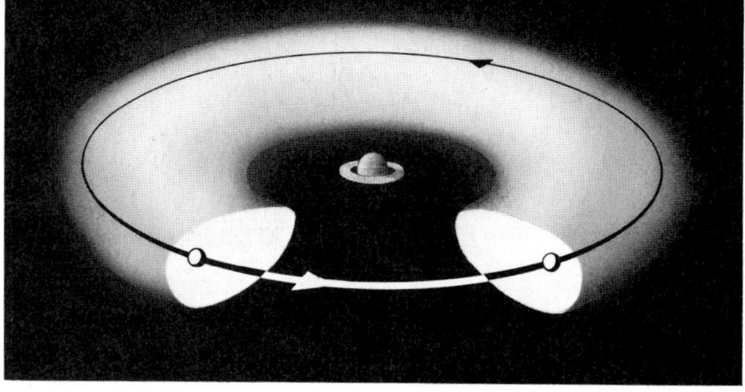

4. Cross-section through Titan
Titan's layered atmosphere is composed of an outer bluish, translucent haze of organic (that is, carbon-based) compounds. The polar hoods are believed to be concentrations of additional compounds within this layer. Beneath it is a relatively clear zone and then a thick layer of smog. This again may be composed of organic compounds that are characteristically red in color. This obscures the view below, but methane clouds may rain or sleet onto the surface. Titan's surface is still a matter for conjecture. It could be solid or liquid, flat or hilly. Methane may well play the same role on Titan as water does on Earth, forming oceans like those on Earth.

5. Torus of neutral hydrogen
A thick disc of neutral hydrogen atoms surrounds Saturn: this extends from the orbit of Rhea to just beyond Titan, giving a width of about 1,000,000 km. The hydrogen supply for this torus is believed to come from Titan. Atoms are thought to escape from the upper atmosphere of the satellite at a rate of about 1 kg/sec to maintain the observed hydrogen density. The torus, which may also be composed of oxygen, co-rotates with the magnetosphere of Saturn. Titan has no magnetic field of its own but interacts with that of the parent planet. The magnetosphere flows around the satellite, creating electrical currents and waves in the surrounding plasma.

Titan II

Titan's surface

Undoubtedly the surface of Titan is both interesting and varied. The temperature is near the triple point of methane; that is to say, methane may exist either as a liquid, a solid or a vapor, as with water on Earth. There may be a steady rain of methane ice or snow; there could be cliffs of solid methane; there may be oceans or rivers of methane, which in the coldest parts may remain rather slushy. The surface temperature is thought to be about 95 K. Measurements from Voyager 1 suggest that this varies by about three degrees only between equator and poles.

The surface itself cannot be observed because it is hidden by the dense layers of haze lying high above the main cloud layers. It is thought that the smog layers are produced by photochemical processes involving methane and nitrogen. Bombardment by energetic particles from Saturn's magnetosphere may also provide an additional source of energy. Titan itself does not possess an intrinsic magnetic field or magnetosphere which would protect it from this bombardment, but it is significant that the orbit lies close to Saturn's magnetosphere, which is very sensitive to the solar wind pressure. Titan sometimes lies within the magnetosphere of Saturn (as it did at the time of the Voyager 1 pass) and sometimes outside (as it did at the time of the Voyager 2 encounter). When Titan is outside the magnetosphere it is exposed to direct interaction with the solar wind particles, and this may possibly affect the chemistry of the upper atmosphere of the satellite.

Weather systems

Although no small-scale structures have been observed in the clouds of Titan, the atmosphere is likely to possess varied weather systems. In the upper atmosphere, where there are temperature contrasts of about 20 K, there may be winds as strong as 100 ms^{-1}. In fact, the rotation of the upper atmosphere is much more rapid than at lower layers—as on Venus. The circulation of Titan's atmosphere suggests large-scale features in the troposphere, and an inter-hemispherical circulation in the stratosphere.

Observations with ground-based telescopes made earlier in the century showed features on Titan that might be interpreted as large-scale cloud structures, and the fact that nothing of the kind was detected by the Voyagers led to the suggestion that the disc may have been anomalously blank during the Voyager period. On the whole this does not seem likely, because the visual observations were too uncertain. When straining to detect details upon a disc as small as Titan's, it is only too easy to be misled. Unfortunately we can hardly hope to learn a great deal about the actual surface without using a probe carrying radar equipment.

Titan's surface
Conditions on Titan's surface are probably gloomy. Its distance from the Sun and the obscuring aerosol layer and methane clouds mean that even at noon it is no lighter than a moonlit night on Earth. Methane cliffs may rise above a methane sea or lake, and water ice may be exposed in places on the surface. Nitrogen drizzle may fall constantly.

Hyperion

Hyperion is one of the smaller satellites of Saturn. Its elliptical orbit lies between those of Titan and Iapetus, but considerably closer to that of Titan. Hyperion was not well surveyed from Voyager 1, which passed by it at a distance of over 880,000 km, but better results were obtained from Voyager 2 at only 470,840 km.

Like all the other satellites, Hyperion provided its quota of surprises. In particular, it is not spherical but irregular in shape, measuring approximately 400 by 250 by 240 km, with angular features and facets as well as rounded features. A body of this size should normally be regular in form, and the strange figure of Hyperion indicates that something unusual may have happened to it in the remote past. Moreover, the long axis is not pointed towards Saturn, as would be expected in a stable configuration; it is at a definite angle. The obvious inference is that Hyperion has suffered collision with another body and has been knocked out of alignment. Eventually it would return to a stable configuration—longest axis turned Saturnward. If this theory is correct, the collision happened a relatively short time ago. Hyperion is a long way from Saturn (almost 1,500,000 km) and tidal interactions are weak, so it would take a long time for a state of equilibrium to be reestablished. Yet what happened to the colliding body? It has been proposed that Hyperion is itself the remnant of a larger satellite that was disrupted at the time of the collision. The rest may have been broken up.

Hyperion has a lower albedo (0.3) than most of the other satellites, and this has been held to be due to material dusted onto it from the surface of the outermost satellite, Phoebe; but if this were so, then the leading hemisphere of Hyperion would be darker than the trailing hemisphere, as with Iapetus, and this does not seem to be the case. There are, however, marked albedo variations (10 to 20 percent) over the whole of the surface. Another factor not yet known is whether the rotation of Hyperion is synchronous.

Craters and scarps

Several fairly large craters are to be seen on Hyperion. One of them is at least 120 km in diameter, with about 10 km of relief. There are several other comparatively deep craters of the order of 40 to 50 km in diameter, and the surface is dotted with smaller craters with diameters of 10 km or less. However, the most prominent surface features are the scarps, which are linked to form one long scarp system nearly 300 km in length. It marks the boundary of a feature more than 200 km across which has a broad, rather low, dome-like structure in its center, and may be a crater. Some of the scarps may be as much as 30 km above the mean surface level of the satellite.

Hyperion is not identical in nature to the inner icy satellites, but from its low density it is thought that ice makes up a significant fraction of the composition. Moreover, lines due to water ice have been observed in its reflection spectrum. The relatively low albedo and incomplete ice cover may be the result of extrinsic dark rocky material. If the dark material on the leading side of Iapetus is derived in part from Phoebe, then some of this material could also reach Hyperion and darken and redden the surface. Titan would effectively eliminate any dust ejected by Phoebe that was not swept up by Iapetus or Hyperion but spiralled in further towards Saturn. Thus the surface of the satellites interior to Titan would not be contaminated by the dark material which may affect the surfaces of Iapetus and Hyperion. This method of darkening would require the rotation of Hyperion not to be tidally locked to Saturn, otherwise it should show the same leading–trailing asymmetry of the darkened surface as there is on Iapetus.

Images of Hyperion
Voyager 2 images of Hyperion show its irregular shape and areas of light and dark. Impact craters are also clearly visible: their diameters ranging from 120 km (with 10 km of relief) to deep craters with diameters of 40 to 50 km and small features 10 km across. The most prominent features are a series of scarps, some of which are linked.

Map of Iapetus

Baligant	225°W,15°N
Basan	197°W,30°N
Charlemagne	266°W,54°N
Geboin	175°W,56°N
Grandoyne	215°W,18°N
Hamon	271°W,10°N
Marsilion	177°W,41°N
Ogier	274°W,42°N
Othon	344°W,24°N
Turpin	0°,43°N

Cassini Regio	210–340°W,48°S–55°N
Roncevaux Terra	130–300°W,30°S–90°N

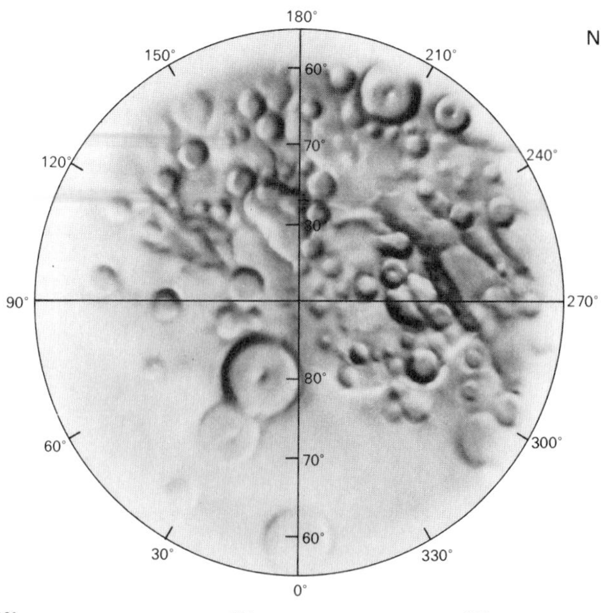

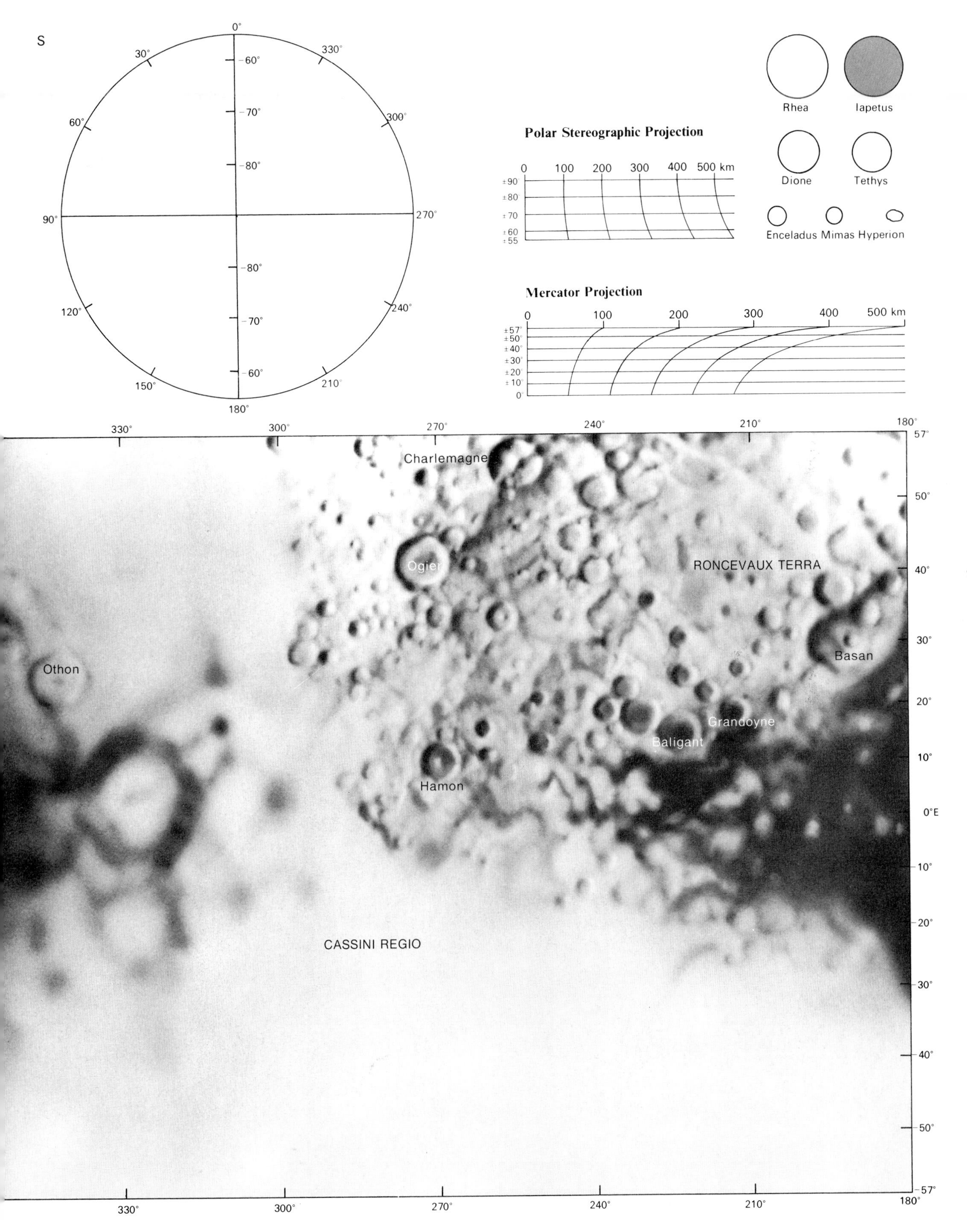
S
0°
30°
330°
60°
300°
90°
270°
120°
240°
150°
210°
180°
60°
70°
80°
Polar Stereographic Projection
0
100
200
300
400
500 km
±90
±80
±70
±60
±55
Mercator Projection
±57°
±50°
±40°
±30°
±20°
±10°
Rhea
Iapetus
Dione
Tethys
Enceladus
Mimas
Hyperion
Charlemagne
Ogier
RONCEVAUX TERRA
Basan
Othon
Grandoyne
Baligant
Hamon
CASSINI REGIO
0°E
57°
50°
40°
30°
20°
10°
-57°

Iapetus

Iapetus is the outermost of Saturn's larger satellites, moving around the planet at a mean distance of almost 3,560,000 km in the comparatively long period of 79 days, and in an orbit that is appreciably inclined (by 14.7°) to the ring-plane. The diameter is 1,440 km, so that in size it is almost a twin of Rhea, but its density is lower—only slightly greater than that of water—and, therefore, so is its mass. Voyager 1 went nowhere near Iapetus, but Voyager 2 passed by at a distance of 909,000 km, which was close enough to obtain reasonable images, though the surface of Iapetus is not nearly as well known as those of the other icy satellites.

1

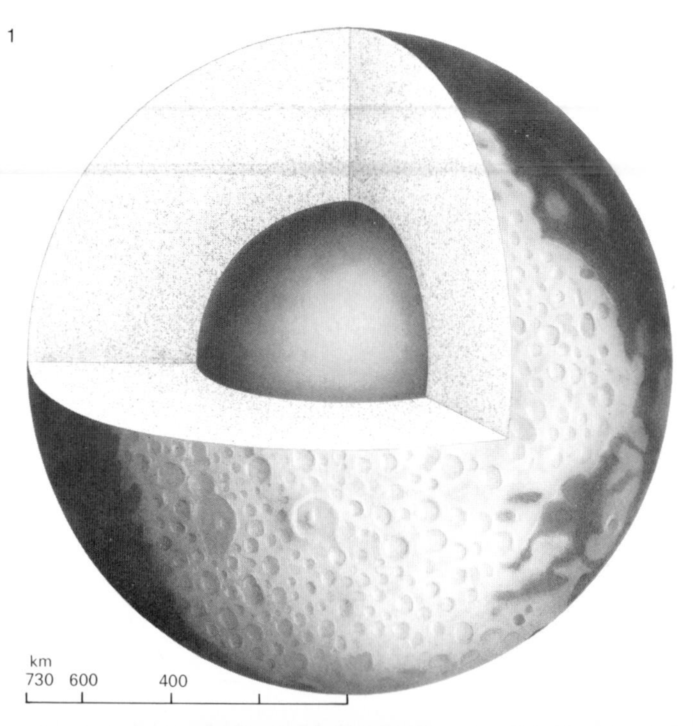

Light and dark sides

Iapetus is variable in brightness. This was established by G. D. Cassini, who discovered the satellite in 1671. Since it was always accepted that the rotation is synchronous, it followed that either the shape of the globe must be irregular or that the two hemispheres were of unequal albedo. When west of Saturn as seen from Earth, Iapetus is an easy telescopic object; when to the east of Saturn it is markedly fainter.

The Voyager results have shown that the second idea is correct. The leading hemisphere of Iapetus is dark (albedo 0.04–0.05), whereas the trailing edge is bright (albedo 0.5). The demarcation line between the two is not abrupt; there is a transition region with a width of between 200 and 300 km, and the boundary itself is somewhat meandering. It is thought that craters occur in both types of terrain, though unfortunately knowledge about the dark leading hemisphere is incomplete.

The reason for this curious division is still a matter for debate. It is certain that the satellite itself is icy, so that it is the dark material that covers an essentially bright surface rather than vice versa. One suggestion is that the leading hemisphere of Iapetus may have been covered by dark material knocked off the surface of Saturn's outermost satellite, Phoebe, by micrometeoritic impacts; the material would spiral downward towards Saturn and be collected by Iapetus. The symmetry of the dark area supports the idea of an exogenous origin. Phoebe is very small, and never approaches within 7,000,000 km of Iapetus; moreover, investigations carried out by D. Cruikshank from the Mauna Kea Observatory in Hawaii, using UKIRT (the United Kingdom Infra-Red Telescope), have shown that the constituent elements of Phoebe differ in nature from the dark material on the leading hemisphere of Iapetus, making the idea of Phoebe as a source rather implausible.

Another objection, however, to the idea of an extraneous cause is that close to the boundary between the bright and dark areas, at approximately latitude 0° and longitude 330°, there is a ring of dark material, 400 km in diameter. Clearly no such ring could have been produced by material "dusted" onto Iapetus' surface.The nature of this strange feature is still unknown. Also of significance is the fact that some of the craters in the bright trailing hemisphere have dark floors. Some are near the center of the trailing hemisphere, where they would be shielded from impact. This would point to an internal origin of the dark material. There is no evidence of craters on the dark side.

It seems likely that the dark material has extruded from Iapetus' interior. This occurred in the case of the Earth's moon, but the material in that case was lava, and no volcanism of this kind is possible upon an icy world such as Iapetus. The extruded material could be a mixture that includes ammonia, soft ice and a dark substance of some kind or other. It has even been suggested that it might be organic in nature. It is not known how thick the deposit may be; it could be a surface coating a millimeter or two deep, or it could be quite thick. It would seem to be a deposit of some kind and, therefore, younger than the cratered terrain. Both regions are red in color, but there is considerable variation from region to region. The dark red material is similar to that on Callisto in Jupiter's system.

3

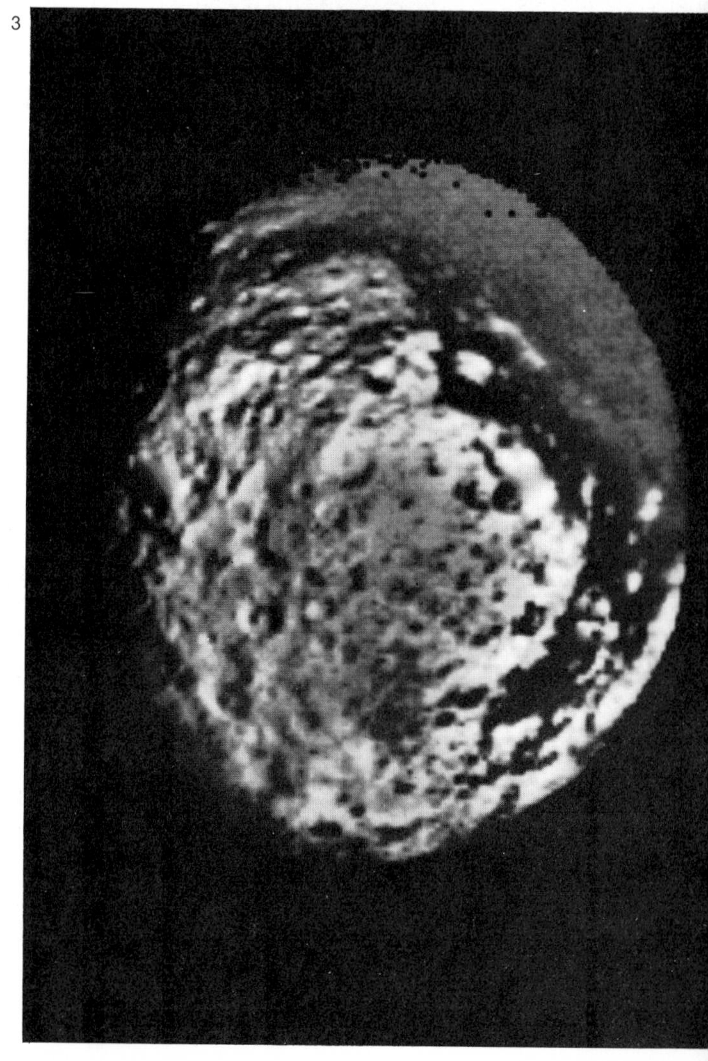

1. Interior of Iapetus
Iapetus has a density of about 1.1, so it is almost as low as that of pure water ice, and the body is at least in part covered with ice. There has been considerable speculation about the nature of the dark material and its origin. The low albedo and red color suggest carbonaceous material such as that found in meteorites and asteroids. This may have erupted from the interior in a slurry with ice and ammonia, and may indicate that Iapetus has a differentiated core.

3. Bright/dark boundary
A sequence of Voyager 2 images of Iapetus shows the anti-Saturn face of the satellite and the boundary between the bright and dark regions. The dark material of Iapetus' leading hemisphere, which reflects only 4 to 5 percent of the light falling on it, extends into the bright trailing hemisphere (with an albedo of nearly 50 percent) in the region of the equator. The north pole is near the large crater on the terminator. There also appears to be dark material on crater floors in the bright zone close to the boundary. Such a sharp but complex boundary tends to refute the idea that the dark material originated somewhere out in space and fell to coat Iapetus, and tends to suggest rather that it extruded from the satellite's interior. Neither theory can be confirmed from Voyager data. These images were taken from a range of 1.1 million km and the smallest features that can be seen are about 20 km in diameter.

2. The dark ring
Voyager 1 only passed within 3.2 million km of Iapetus, but this image of the Saturn-facing hemisphere of the satellite still shows up the dramatic contrast between light and dark regions. Of particular interest is a dark ring extending beyond the dark hemisphere. This feature is approximately 400 km in diameter and has a dark spot in its center. It is probably an impact structure, outlined by dark material that was thrown out from the point where the impacting body struck.

2

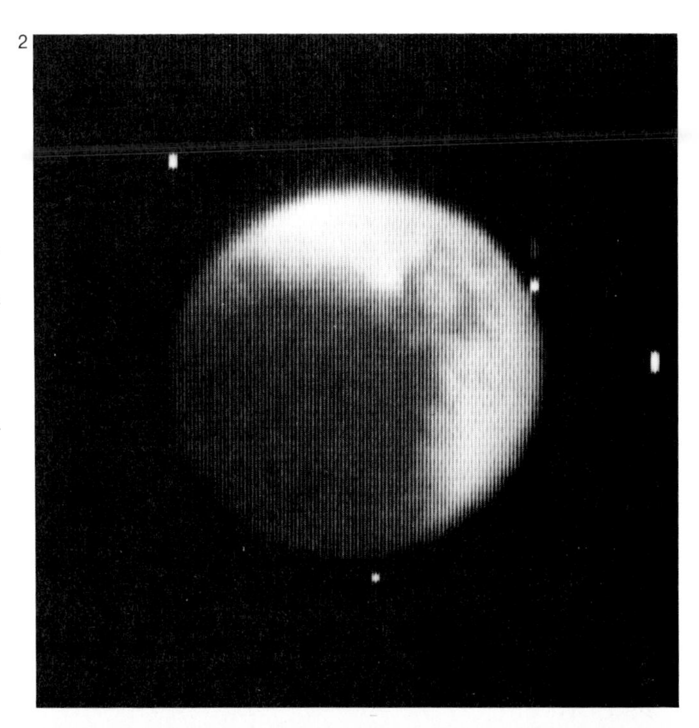

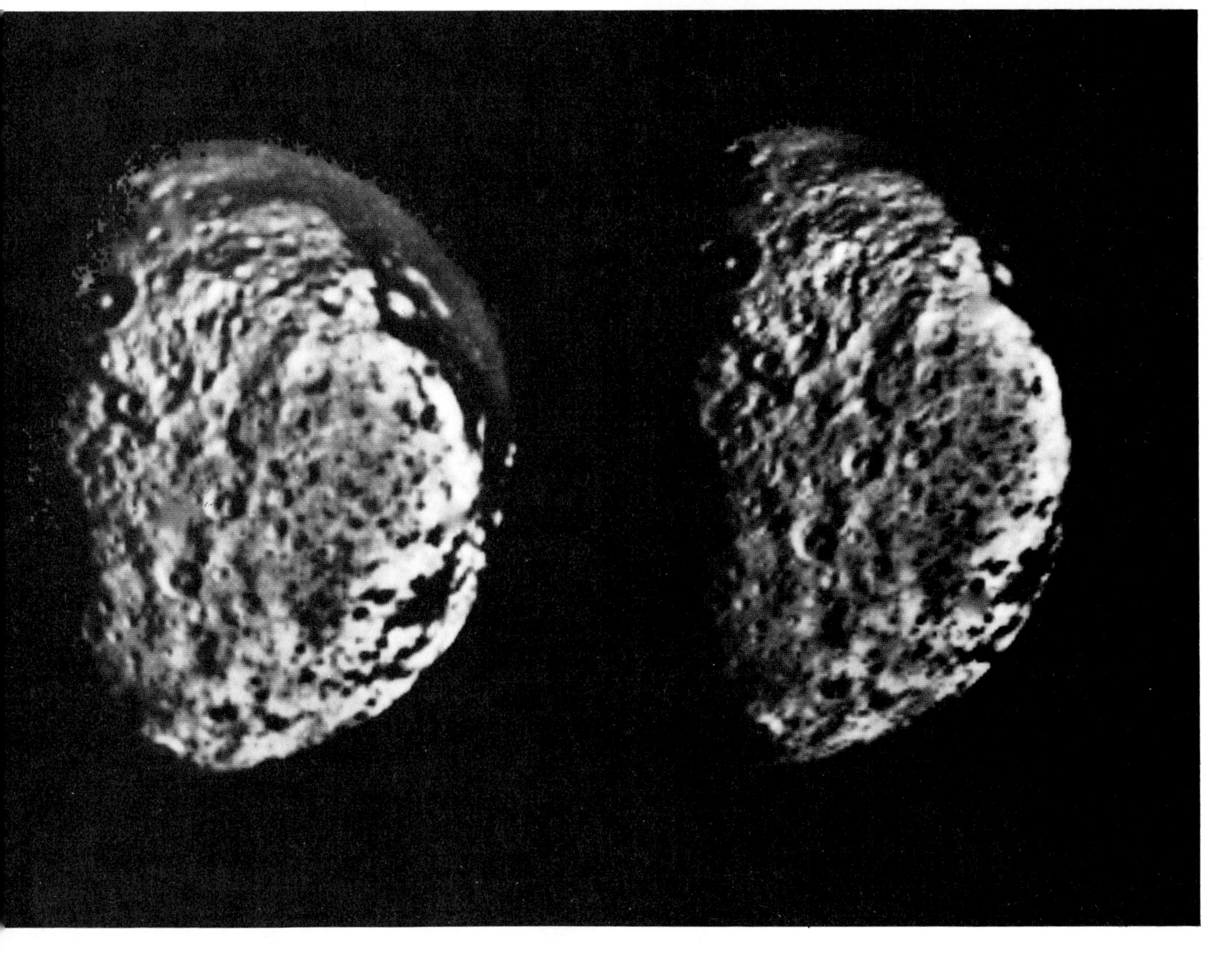

Phoebe

The outermost of Saturn's satellite system, Phoebe, was discovered in 1898 and was the first satellite discovery to be made with the aid of photography. Phoebe is smaller than any of the other named satellites, with a diameter of about 160 km, so it is a very faint telescopic object as seen from Earth. Its mean distance from Saturn is more than 10,000,000 km, and it moves in a retrograde orbit with a period of 550.4 days.

Phoebe—alone of the named satellites—was not well surveyed by either Voyager. The minimum distance from Voyager 1 was more than 13,500,000 km, and Voyager 2 came no closer to it than 1,473,000 km. Therefore we have no large-scale pictures of the surface, and Phoebe is less well known than any other member of the Saturnian family. This is unfortunate, because it is unlike any of the remaining satellites, and may be a captured asteroid.

The surface is darkish (with an albedo of about 0.05), but less red than that of Iapetus. The low albedo and color suggest that Phoebe is similar to a class of asteroids that is believed to be common in the outer Solar System. These asteroids are thought to be of primitive composition. This could mean that Phoebe is the first relatively unmodified primitive body in the outer Solar System to be photographed by a spacecraft. A few discrete features can be made out, but nothing definite. The albedo contrasts of these features are as much as 50 percent. It is reasonable to assume that Phoebe is cratered, but proof must await the findings of a future spaceprobe encounter.

Phoebe, unlike Hyperion, is spherical in form as far as we can tell. From the identified features, it seems that the rotation period is of the order of 9 hr, so Phoebe does not have synchronous rotation, which again may make it unique. More important is the fact that it has a retrograde motion, and the orbit is inclined at 150° to Saturn's equatorial plane.

Origin of Phoebe

These characteristics have led to the suggestion that Phoebe is not a bona-fide satellite at all, but was captured by Saturn in the remote past, and has not changed since it accreted early in the evolution of the Solar System. It may have been ejected from the inner Solar System by the gravitational field of the growing planet Jupiter. If Phoebe is less heavily cratered than the inner icy satellites, it may be assumed that capture by Saturn took place after the last bombardment of Iapetus, otherwise Phoebe would have been in grave danger of disruption. This, in turn, assumes that craters on the other satellites are indeed mainly of impact origin. Less plausible is the suggestion that Phoebe represents the encrusted nucleus of a gigantic comet. It seems that the mass of Phoebe, slight though it may be by Solar System standards, is too great for the satellite to be cometary in nature.

The Chiron enigma

The theory that Phoebe is a captured body, formerly moving around the Sun in an independent path, seems less unlikely now than it did a few years ago, because there is positive evidence that asteroidal-type objects do move in this part of the Solar System. The first important discovery was made in 1977 by Charles Kowal, using the Schmidt telescope at the Palomar Observatory in California. During a search for distant comets, Kowal discovered a remarkable object which was found to be moving in an orbit for the most part between those of Saturn and Uranus. It was not a comet, and was assumed to be an asteroid, though the region beyond Saturn is not a region where asteroids are expected to be found. It has been named Chiron, and given an official asteroid number, 2060. At perihelion its distance from the Sun is 8.5 astronomical units or 1,278 million km, which is just inside Saturn's orbit; at perihelion it recedes to 18.9 astronomical units, or 2,827 million km. Its orbital inclination is 6.9°. The diameter is uncertain because there is no information about the albedo, but it may be about 100 km if its composition is icy. Recent studies show that it is more likely to be rocky than icy, in which case the albedo will be comparatively low, and it may have a diameter of 300–400 km, which is large by asteroidal standards. Chiron will next reach perihelion in 1996.

Chiron's orbit may not be stable, and in future it may be thrown out of the Solar System. Meanwhile it is believed that in 1664 BC Chiron approached Saturn to within a distance of 16,000,000 km, which is not much greater than the distance between Saturn and Phoebe. It is tempting to believe that Phoebe and Chiron are of the same type, in which case Phoebe is a latecomer to the Saturnian family. No doubt other objects of a similar type exist, and it is even suggested that Pluto may be essentially asteroidal in as much as it appears to be smaller and less massive than the Earth's moon, and is presumably icy.

In short, the precise nature of Phoebe is uncertain, but its small size, relatively low albedo, and above all its retrograde motion indicate that it may well be asteroidal.

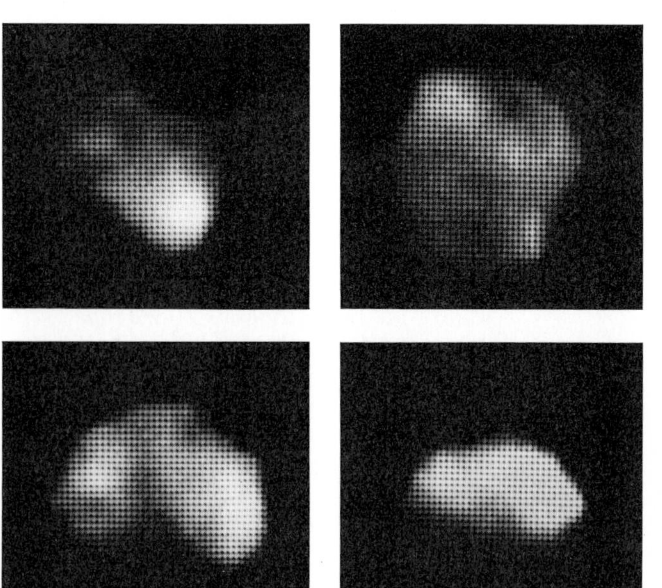

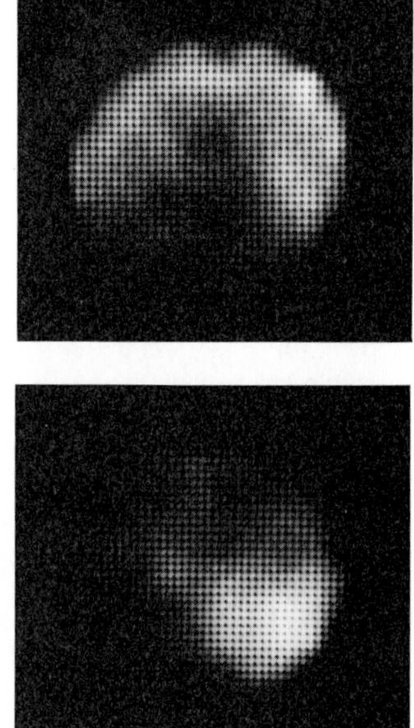

Images of Phoebe
Voyager 1 obtained no data on Saturn's most distant satellite. Voyager 2 took a series of images but still from a distance of more than 2.2 million km. The first five frames were at intervals of 70° of longitude. The final one was taken one revolution after the first frame. Even at this distance bright and dark features, with contrasting albedoes of 50 percent, can be seen clearly. The north pole of Phoebe's rotation is at the top of each image.

Conclusion

Any theory of the origin of the Solar System has to account for a number of known physical properties. It must explain, for example, the distribution of chemical elements throughout the System, and also the distribution of angular momentum (the product of the mass of a spinning particle, its velocity, and the distance from the center of rotation); in more general terms, the theory must explain the extreme differences between the "terrestrial" planets (Mercury, Venus, the Earth and Mars) and the major planets beyond the asteroid belt (Jupiter, Saturn, Uranus and Neptune). The former group consists of relatively small dense bodies, in contrast with the low-density objects with optically thick atmospheres which are typical of the major planets.

The "nebular hypothesis"

The most widely held modern view is known as the "nebular hypothesis", an early version of which was originally proposed in 1755 by Kant and, in a modified form, by Laplace in 1796. Neither of these early versions meets the difficulties of the problem, but more recent developments of the hypothesis succeed in accounting for at least some of the known facts. According to this theory, the Sun and stars have condensed out of clouds of interstellar dust and gas. This interstellar haze was probably thinly spread, until some sudden neighboring event, such as the explosion of a nearby star, disturbed the equilibrium. The resulting shockwave caused concentrations of particles, which, once started, continued to coalesce under the force of their own gravity. At the same time, the further it contracted, the faster the cloud would have to spin to conserve its angular momentum. However, on its own this mechanism would lead to a situation in which most of the angular momentum of the Solar System would be concentrated at the center, whereas in fact the Sun accounts for only 2 percent of the total. Jupiter, on the other hand, accounts for 60 percent. Other factors, such as radiation and magnetic forces, may have played a part in redistributing the angular momentum in such a way as to produce the results that are observed today.

The composition of material is equally important. In the absence of more detailed evidence, several alternative theories are available. According to one view, for example, the primordial nebula was originally cool, possibly containing some of the organic molecules which are essential to terrestrial life, as well as large proportions of hydrogen, helium, and other chemicals such as carbon monoxide, water and ammonia. The increasing temperature of the spinning disc, however, boiled away the lighter, more volatile elements in the near vicinity of the forming star. Materials with high melting points, such as iron and the silicate materials that make up rocks, were left nearer the star, where they formed dense planets of the "terrestrial" type; the lighter elements condensed further away, so that the outer planets would be expected to contain a greater proportion of hydrogen and helium and would resemble the Sun in terms of composition.

The origin of Saturn, rings and satellites

The planets are currently believed to have formed between 4,500 and 5,000 million years ago by a process of gravitational collapse similar to the process that formed the Sun. Jupiter and Saturn swept up almost all the material available to the planets. At first Jupiter and Saturn, and perhaps Uranus and Neptune, were several hundred times larger than their present size. Under the influence of their own gravitation they gradually shrank, rotating faster, and eventually increasing the centrifugal force associated with the rotation to create a gaseous envelope distinct from the nearby spherical concentration of gas inside which the young planet continued to condense. The envelope served as a source of solid grains from which the system of satellites formed. After an interval of about 25 million years the planetary envelopes disappeared: perhaps they were blown away by a wind of ionized gas from the young Sun, or, alternatively, friction within the envelopes caused them to collapse onto their planets. The result was a planet, rings and a system of moons.

Today, both Saturn and Jupiter give off more energy than they receive from the Sun. For Jupiter this heat can be accounted for by a small gravitational contraction of about 1 millimeter per year. In the case of Saturn this does not seem to be the only mechanism involved, and a clue about an additional source of heat comes from the fact that the planet is depleted in helium compared with Jupiter and the Sun. This depletion is thought to be due to processes taking place in the interior of the planet where droplets of helium fall as rain through the fluid hydrogen layers towards the planet's core.

The processes related to the excess heat emitted by both Jupiter and Saturn have been taking place throughout their period of evolution. At the time their moons were forming the planets were contracting much more rapidly so that they probably radiated considerably more heat than is detected today. This heat would certainly have affected and even controlled the temperature and thus the composition of the solid matter in the planetary envelopes. At a given time the temperature would have increased towards the planet, and at a given place the envelope would have steadily cooled with time. These characteristics would affect not only the composition of the ring particles, but also the composition of the moons in orbit around the parent planet.

Jupiter's star-like early stages would have produced enough heat to make the environment of the inner satellites significantly warmer than the outer members. During some 100 million years this would have permitted the formation and retention of water ice at the present positions of Ganymede and Callisto. As a result there is a general decrease in the density of the Galilean satellites outward from Jupiter. Saturn seems to be quite different. The early luminosity of Saturn was weaker than Jupiter's owing to the smaller mass of Saturn. Thus, even near to the planet, no water was driven out of the Saturnian system. Instead we find systems of water-ice satellites and rings. Titan is the exception in the Saturnian system of satellites, and possibly in the Solar System in general, in that it has an atmosphere. The temperature at its surface is high enough to maintain significant amounts of methane and ammonia gas in equilibrium with their corresponding ices. Triton, the giant moon of Neptune, may simply be too cold for this to occur. Also, Titan formed from material that could generate an atmosphere, namely ices containing ammonia and methane. The Galilean satellites formed in the hot nebula of Jupiter and probably lack significant quantities of these materials.

Future Saturn probes

At the moment no further probes to Saturn are planned, and there is no longer the chance of making use of the situation which prevailed in the late 1970s, when the arrangement of the four giant planets made it possible to launch a multiplanet probe such as Voyager 2. However, it does not seem likely that many decades will pass by before further attempts are made to explore Saturn.

Titan is, in its way, as important and as interesting as Saturn itself. Sending another Voyager-type probe to Titan would be of little value, because it is known that the satellite's clouds are opaque. What would be required is a probe that would enter a closed orbit around Titan and map the surface by means of radar, as Pioneer 13 did for the cloud-covered planet Venus. Landing probes could also be sent to Titan, and there is no reason why they should not continue to operate for long periods after arrival, but there is a chance that such a probe could come down in a Titanian ocean of liquid methane, so it would have to be designed to cope with such an eventuality. The chances of finding any life on Titan are extremely small; the basic materials for life exist there, but temperatures are extremely low. Certainly an orbiting Titan probe would be of immense importance.

Tables

The tables presented on this and the opposite page are intended mainly for the practical observer of Saturn (*see* page 92). Specialist publications such as the *Astronomical Ephemeris* and the *American Ephemeris* are published annually and will, of course, remain the primary sources of information for scientific research: but it is hoped that the information provided here, devoted specifically to the planet Saturn, brings together the most critical data in a form that will be useful as general reference or, perhaps, will be of use to the observer when more detailed works are not readily available.

Figure 1

The diagram shows the orbital elements of any planet in the Solar System on its journey around the Sun. The first part of the diagram shows the basic dimensions of an elliptical orbit, while the second describes the elements of a planet's orbit with reference to the orbit of the Earth. Table 1 gives the dates on which Saturn is in opposition (*see* page 6), and is thus best placed for observation, for the period from 1983 to the end of the century. The table also gives the magnitude of Saturn for each opposition. The planet was at its brightest during the 1970s, but from the data it can be seen that it starts to get brighter again towards the turn of the century. The figures in the table were obtained from the Journal of the British Astronomical Association.

The rotation periods for Saturn are also given on this page. In the case of the planet Jupiter there are tables that give the agreed longitudes of System I (equatorial) and System II (the rest of the planet) for regular intervals, and these have been in use for many years. For Saturn, however, this information is much less easy to determine and, although such tables are available for Saturn, they must be regarded as somewhat arbitrary. It is mainly for this reason that we have not included more detailed data for Saturn's rotation in this book. The figures for System I and System II in this table are pre-Voyager, Earth-based figures and pre-date an official System designation. System III is the solid-body rotational period, determined by radio signals.

Tables 2 and 3

Tables 2 and 3 give the fundamental data for each planetary member of the Solar System. Table 2 gives orbital data for each planet, and Table 3 gives physical data (including figures for the Sun and the Moon). Table 2 may be read in conjunction with diagram 1 which, in fact, illustrates the elements included in the table. It should be noted, however, that the longitude figures listed in the table below represent mean values that have been calculated, rather than the "true" values.

Table 4

This table sets out the Right Ascension and Declination of Saturn for the period from 1982 to 1992. Ephemeris data is not normally published for more than four or five years in advance, and the accuracy of the *Astronomical Ephemeris* is only achieved for the current year. The reason for this is that considerable uncertainties exist over projections further into the future.

The data in Table 4 is reproduced by courtesy of H. M. Nautical Almanac Office, who kindly supplied the computer print-out from which the figures were taken.

1A
Planet
ν
a
ae
Sun
Perihelion

B
Ecliptic
Planet
Desc.Node
ν
Perihelion
ω
i
Ω
Asc.Node
♈

1. Orbital elements
Planets travel in elliptical orbits with the Sun at one focus. The dimensions of the ellipse may be described by two elements, the semi-major axis (a), and the eccentricity (e) as defined by diagram (**A**). Another four elements are needed to specify fully the orbit of a planet as illustrated in diagram (**B**), where the plane of the ecliptic is defined by the Earth's orbit. The angle i is the inclination of the orbit to the plane of the ecliptic; Ω is the longitude of the ascending node; ω is the "argument" of the perihelion; and L is the longitude of the planet at a specified moment. (L is given by $\Omega + \omega + \nu$.) Finally, there is the period (T).

Rotational period of Saturn
Equator (System I): 10 hr 15 min
Mid-latitude (II): 10 hr 38 min
System III: 10 hr 39.4 min

Table 1: Oppositions 1983–2000

Date	Mag.	Date	Mag.
21.4.83	+0.4	19. 8.93	+0.5
3.5.84	+0.3	1. 9.94	+0.7
15.5.85	+0.2	14. 9.95	+0.8
27.5.86	+0.2	26. 9.96	+0.7
9.6.87	+0.2	10.10.97	+0.4
20.6.88	+0.2	23.10.98	+0.2
2.7.89	+0.2	6.11.99	0.0
14.7.90	+0.3	19.11.2000	−0.1
26.7.91	+0.3		
7.8.92	+0.4		

Table 2: Mean elements of the planetary orbits for epoch 1980 Jan. 1.5 E.T.

Planet	Mean distance A.U.	Mean distance a millions of km	Eccentricity e	Inclination to ecliptic i ° ′ ″	Mean longitude of asc. node Ω ° ′ ″	Mean longitude of perihelion ϖ $(=\omega+\Omega)$ ° ′ ″	Mean longitude at the epoch L ° ′ ″	Sidereal period days
Mercury	0.3870987	57.91	0.2056306	7 00 15.7	48 05 39.2	77 08 39.4	237 26 09.2	87.969
Venus	0.7233322	108.21	0.0067826	3 23 40.0	76 29 59.2	131 17 22.7	358 08 12.4	224.701
Earth	1.0000000	149.60	0.0167175	— — —	— — —	102 35 47.2	100 18 43.2	365.256
Mars	1.5236915	227.94	0.0933865	1 50 59.3	49 24 11.6	335 41 27.2	127 06 26.0	686.980
Jupiter	5.2028039	778.34	0.0484681	1 18 15.2	100 14 48.0	14 00 01.9	147 05 29.8	4332.59
Saturn	9.5388437	1,427.01	0.0556125	2 29 20.9	113 28 55.4	92 39 22.9	165 22 24.3	10,759.20
Uranus	19.181826	2,869.6	0.0472639	0 46 23.5	73 53 54.2	170 20 10.7	227 17 14.5	30,684.8
Neptune	30.058021	4,496.7	0.0085904	1 46 18.8	131 33 36.7	44 27 01.1	260 54 42.6	60,190.5
Pluto	39.44	5,900.0	0.250	17 12 00	110	223		90,465.0

Table 3: Physical data for the Sun, Moon and planets

Name	Diameter equatorial km	Diameter polar km	Inclination degrees	Reciprocal mass (Sun=1)	Mass kg	Density (water=1)	Escape velocity km s^{-1}	Volume (Earth=1)	Surface gravity (Earth=1)	Mean vis. opposition Mag.	Albedo
Sun	1,392,530	1,392,530	7.25	1	1.9891×10^{30}	1.41	617.3	1.3×10^{6}	28.0	−26.8	—
Moon	3,476	3,476	1.53	27,068,000	7.3483×10^{22}	3.34	2.37	0.02	0.165	−12.7	0.07
Mercury	4,878	4,878	0	6,023,600	3.3022×10^{23}	5.43	4.25	0.06	0.377	0.0	0.06
Venus	12,104	12,104	178	408,523.5	4.8689×10^{24}	5.24	10.36	0.86	0.902	−4.4	0.76
Earth	12,756	12,714	23.44	328,900.5	5.9742×10^{24}	5.52	11.18	1.00	1.000	—	0.36
Mars	6,794	6,759	25.20	3,098,710	6.4191×10^{23}	3.93	5.02	0.15	0.379	−2.0	0.16
Jupiter	142,800	134,200	3.12	1,047.355	1.899×10^{27}	1.32	59.6	1,323	2.69	−2.6	0.73
Saturn	120,000	108,000	26.73	3,498.5	5.684×10^{26}	0.70	35.6	752	1.19	+0.7	0.76
Uranus	52,000	49,000	97.86	22,869	8.6978×10^{25}	1.25	21.1	64	0.93	+5.5	0.93
Neptune	48,400	47,400	29.56	19,314	1.028×10^{26}	1.77	24.6	54	1.22	+7.8	0.62
Pluto	3,000	3,000	90	3,000,000	6.6×10^{23}	4.7	7.7	0.01	0.20	+14.9	0.5

Table 4: Position of Saturn 1982–1992

Date	R.A.	Dec	Date	R.A.	Dec	Date	R.A.	Dec	Date	R.A.	Dec	Date	R.A.	Dec
d mo yr	h m s	° ′ ″	d mo yr	h m s	° ′ ″	d mo yr	h m s	° ′ ″	d mo yr	h m s	° ′ ″	d mo yr	h m s	° ′ ″
28.10.82	13 40 46	− 8 2 20	27. 9.84	14 48 29	−14 0 32	28. 8.86	16 6 57	−19 7 43	28. 7.88	17 46 8	−22 20 20	28. 6.90	19 40 13	−21 20 41
7.11.82	13 45 18	− 8 27 38	7.10.84	14 52 35	−14 20 17	7. 9.86	16 8 38	−19 14 32	7. 8.88	17 44 16	−22 21 13	8. 7.90	19 37 12	−21 28 37
17.11.82	13 49 43	− 8 51 39	17.10.84	14 56 58	−14 40 33	17. 9.86	16 10 55	−19 22 46	17. 8.88	17 43 1	−22 22 22	18. 7.90	19 34 5	−21 36 38
27.11.82	13 53 56	− 9 13 59	27.10.84	15 1 52	−15 0 60	27. 9.86	16 13 46	−19 32 11	27. 8.88	17 42 26	−22 23 49	28. 7.90	19 31 0	−21 44 19
7.12.82	13 57 53	− 9 34 14	6.11.84	15 6 15	−15 21 15	7.10.86	16 17 8	−19 42 31	6. 9.88	17 42 32	−22 25 34	7. 8.90	19 28 8	−21 51 22
17.12.82	14 1 29	− 9 52 5	16.11.84	15 11 2	−15 40 57	17.10.86	16 20 57	−19 53 28	16. 9.88	17 43 21	−22 27 35	17. 8.90	19 25 36	−21 57 30
27.12.82	14 4 41	−10 7 11	26.11.84	15 15 47	−15 59 49	27.10.86	16 25 10	−20 4 44	26. 9.88	17 44 51	−22 29 48	27. 8.90	19 23 32	−22 2 32
6. 1.83	14 7 23	−10 19 18	6.12.84	15 20 27	−16 17 32	6.11.86	16 29 41	−20 16 4	6.10.88	17 47 0	−22 32 6	6. 9.90	19 22 3	−22 6 19
16. 1.83	14 9 32	−10 28 10	16.12.84	15 24 57	−16 33 51	16.11.86	16 34 27	−20 27 11	16.10.88	17 49 47	−22 34 21	16. 9.90	19 21 13	−22 8 48
26. 1.83	14 11 5	−10 33 37	26.12.84	15 29 12	−16 48 32	26.11.86	16 39 22	−20 37 53	26.10.88	17 53 6	−22 36 25	26. 9.90	19 21 4	−22 9 53
5. 2.83	14 11 59	−10 35 36	**5. 1.85**	15 33 7	−17 1 21	6.12.86	16 44 23	−20 47 55	5.11.88	17 56 55	−22 38 10	6.10.90	19 21 37	−22 9 35
15. 2.83	14 12 12	−10 34 3	15. 1.85	15 36 37	−17 12 12	16.12.86	16 49 24	−20 57 10	15.11.88	18 1 8	−22 39 26	16.10.90	19 22 53	−22 7 53
25. 2.83	14 11 46	−10 29 7	25. 1.85	15 39 38	−17 20 54	26.12.86	16 54 20	−21 5 28	25.11.88	18 5 42	−22 40 8	26.10.90	19 24 49	−22 4 45
7. 3.83	14 10 40	−10 21 2	4. 2.85	15 42 5	−17 27 23	**5. 1.87**	16 59 6	−21 12 45	5.12.88	18 10 32	−22 40 9	5.11.90	19 27 22	−22 0 12
17. 3.83	14 9 0	−10 10 9	14. 2.85	15 43 55	−17 31 36	15. 1.87	17 3 37	−21 18 56	15.12.88	18 15 33	−22 39 25	15.11.90	19 30 30	−21 54 17
27. 3.83	14 6 50	− 9 56 59	24. 2.85	15 45 6	−17 33 30	25. 1.87	17 7 48	−21 24 4	25.12.88	18 20 39	−22 37 56	25.11.90	19 34 8	−21 46 58
6. 4.83	14 4 17	− 9 42 13	6. 3.85	15 45 35	−17 33 9	4. 2.87	17 11 35	−21 28 7	**4. 1.89**	18 25 46	−22 35 44	5.12.90	19 38 11	−21 38 21
16. 4.83	14 1 29	− 9 26 33	16. 3.85	15 45 23	−17 30 36	14. 2.87	17 14 52	−21 31 9	14. 1.89	18 30 49	−22 32 51	15.12.90	19 42 36	−21 28 30
26. 4.83	13 58 36	− 9 10 50	26. 3.85	15 44 29	−17 25 59	24. 2.87	17 17 36	−21 33 14	24. 1.89	18 35 42	−22 29 24	25.12.90	19 47 17	−21 17 30
6. 5.83	13 55 45	− 8 55 54	5. 4.85	15 42 58	−17 19 31	6. 3.87	17 19 42	−21 34 26	3. 2.89	18 40 21	−22 25 32	**4. 1.91**	19 52 9	−21 5 30
16. 5.83	13 53 7	− 8 42 30	15. 4.85	15 40 53	−17 11 29	16. 3.87	17 21 9	−21 34 53	13. 2.89	18 44 40	−22 21 25	14. 1.91	19 57 8	−20 52 40
26. 5.83	13 50 48	− 8 31 22	25. 4.85	15 38 22	−17 2 12	26. 3.87	17 21 53	−21 34 36	23. 2.89	18 48 35	−22 17 16	24. 1.91	20 2 7	−20 39 13
5. 6.83	13 48 56	− 8 23 2	5. 5.85	15 35 32	−16 52 8	5. 4.87	17 21 54	−21 33 42	5. 3.89	18 52 3	−22 13 16	3. 2.91	20 7 4	−20 25 24
15. 6.83	13 47 35	− 8 17 56	15. 5.85	15 32 32	−16 41 46	15. 4.87	17 21 13	−21 32 14	15. 3.89	18 54 57	−22 9 38	13. 2.91	20 11 52	−20 11 29
25. 6.83	13 46 49	− 8 16 19	25. 5.85	15 29 30	−16 31 37	25. 4.87	17 19 52	−21 30 16	25. 3.89	18 57 16	−22 6 38	23. 2.91	20 16 26	−19 57 46
5. 7.83	13 46 40	− 8 18 16	4. 6.85	15 26 37	−16 22 16	5. 5.87	17 17 55	−21 27 50	4. 4.89	18 58 56	−22 4 24	5. 3.91	20 20 44	−19 44 36
15. 7.83	13 47 8	− 8 23 46	14. 6.85	15 24 1	−16 14 13	15. 5.87	17 15 28	−21 25 2	14. 4.89	18 59 55	−22 3 5	15. 3.91	20 24 40	−19 52 17
25. 7.83	13 48 12	− 8 32 39	24. 6.85	15 21 49	−16 7 55	25. 5.87	17 12 37	−21 21 55	24. 4.89	19 0 12	−22 2 47	25. 3.91	20 28 10	−19 21 9
4. 8.83	13 49 52	− 8 44 40	4. 7.85	15 20 8	−16 3 45	4. 6.87	17 9 33	−21 18 39	4. 5.89	18 59 47	−22 3 33	4. 4.91	20 31 10	−19 11 34
14. 8.83	13 52 5	− 8 59 34	14. 7.85	15 19 1	−16 1 57	14. 6.87	17 6 23	−21 15 22	14. 5.89	18 58 41	−22 5 20	14. 4.91	20 33 37	−19 3 48
24. 8.83	13 54 48	− 9 16 58	24. 7.85	15 18 31	−16 2 41	24. 6.87	17 3 18	−21 12 15	24. 5.89	18 56 58	−22 8 1	24. 4.91	20 35 29	−18 58 8
3. 9.83	13 57 59	− 9 36 29	3. 8.85	15 18 41	−16 5 58	4. 7.87	17 0 26	−21 9 33	3. 6.89	18 54 42	−22 11 28	4. 5.91	20 36 43	−18 54 47
13. 9.83	14 1 33	− 9 57 44	13. 8.85	15 19 30	−16 11 42	14. 7.87	16 57 57	−21 7 28	13. 6.89	18 52 0	−22 15 28	14. 5.91	20 37 17	−18 53 52
23. 9.83	14 5 28	−10 20 17	23. 8.85	15 20 56	−16 19 47	24. 7.87	16 55 56	−21 6 11	23. 6.89	18 49 0	−22 19 46	24. 5.91	20 37 10	−18 55 27
3.10.83	14 9 40	−10 43 44	2. 9.85	15 22 59	−16 29 57	3. 8.87	16 54 31	−21 5 53	3. 7.89	18 45 51	−22 24 9	3. 6.91	20 36 25	−18 59 26
13.10.83	14 14 5	−11 7 41	12. 9.85	15 25 36	−16 41 56	13. 8.87	16 53 43	−21 6 40	13. 7.89	18 42 42	−22 28 25	13. 6.91	20 35 3	−19 5 40
23.10.83	14 18 39	−11 31 42	22. 9.85	15 28 44	−16 55 25	23. 8.87	16 53 37	−21 8 34	23. 7.89	18 39 42	−22 32 24	23. 6.91	20 33 7	−19 13 51
2.11.83	14 23 18	−11 55 25	2.10.85	15 32 19	−17 10 4	2. 9.87	16 54 12	−21 11 34	2. 8.89	18 37 2	−22 35 60	3. 7.91	20 30 42	−19 23 34
12.11.83	14 27 57	−12 18 27	12.10.85	15 36 17	−17 25 34	12. 9.87	16 55 27	−21 15 35	12. 8.89	18 34 47	−22 39 7	13. 7.91	20 27 57	−19 34 20
22.11.83	14 32 32	−12 40 24	22.10.85	15 40 36	−17 41 33	22. 9.87	16 57 22	−21 20 29	22. 8.89	18 35 7	−22 41 44	23. 7.91	20 24 58	−19 45 37
2.12.83	14 36 59	−13 0 58	1.11.85	15 45 10	−17 57 41	2.10.87	16 59 54	−21 26 5	1. 9.89	18 32 3	−22 43 50	2. 8.91	20 21 54	−19 56 50
12.12.83	14 41 13	−13 19 50	11.11.85	15 49 56	−18 13 40	12.10.87	17 3 0	−21 32 10	11. 9.89	18 31 43	−22 45 24	12. 8.91	20 18 56	−20 7 27
22.12.83	14 45 9	−13 36 39	21.11.85	15 54 48	−18 29 13	22.10.87	17 6 37	−21 38 33	21. 9.89	18 32 1	−22 46 25	22. 8.91	20 16 12	−20 17 0
1. 1.84	14 48 43	−13 51 13	1.12.85	15 59 42	−18 44 2	1.11.87	17 10 39	−21 44 59	1.10.89	18 33 4	−22 46 51	1. 9.91	20 13 51	−20 25 6
11. 1.84	14 51 50	−14 3 17	11.12.85	16 4 34	−18 57 55	11.11.87	17 15 4	−21 51 15	11.10.89	18 34 48	−22 46 38	11. 9.91	20 11 59	−20 31 26
21. 1.84	14 54 25	−14 12 38	21.12.85	16 9 19	−19 10 39	21.11.87	17 19 46	−21 57 11	21.10.89	18 37 10	−22 45 44	21. 9.91	20 10 42	−20 35 47
31. 1.84	14 56 25	−14 19 10	31.12.85	16 13 50	−19 22 5	1.12.87	17 24 41	−22 2 36	31.10.89	18 40 8	−22 44 5	1.10.91	20 10 5	−20 38 1
10. 2.84	14 57 47	−14 22 45	**10. 1.86**	16 18 5	−19 32 6	11.12.87	17 29 44	−22 7 21	10.11.89	18 43 38	−22 41 36	11.10.91	20 10 9	−20 38 6
20. 2.84	14 58 29	−14 23 24	20. 1.86	16 21 57	−19 40 35	21.12.87	17 34 50	−22 11 23	20.11.89	18 47 36	−22 38 15	21.10.91	20 10 55	−20 35 59
1. 3.84	14 58 30	−14 21 9	30. 1.86	16 25 21	−19 47 30	31.12.87	17 39 54	−22 14 36	30.11.89	18 51 58	−22 33 59	31.10.91	20 12 21	−20 31 43
11. 3.84	14 57 51	−14 16 6	9. 2.86	16 28 15	−19 52 48	**10. 1.88**	17 44 51	−22 17 1	10.12.89	18 56 37	−22 28 48	10.11.91	20 14 27	−20 25 21
21. 3.84	14 56 33	−14 8 31	19. 2.86	16 30 33	−19 56 30	20. 1.88	17 49 35	−22 18 38	20.12.89	19 1 30	−22 22 43	20.11.91	20 17 9	−20 16 57
31. 3.84	14 54 41	−13 58 45	1. 3.86	16 32 12	−19 58 37	30. 1.88	17 54 2	−22 19 33	30.12.89	19 6 32	−22 15 47	30.11.91	20 20 23	−20 6 38
10. 4.84	14 52 20	−13 47 12	11. 3.86	16 33 9	−19 59 12	9. 2.88	17 58 7	−22 19 51	**9. 1.90**	19 11 37	−22 8 7	10.12.91	20 24 5	−19 54 32
20. 4.84	14 49 39	−13 34 28	21. 3.86	16 33 25	−19 58 18	19. 2.88	18 1 46	−22 19 39	19. 1.90	19 16 40	−21 59 51	20.12.91	20 28 11	−19 40 48
30. 4.84	14 46 45	−13 21 10	31. 3.86	16 32 58	−19 55 60	29. 2.88	18 4 53	−22 19 6	29. 1.90	19 21 38	−21 51 9	30.12.91	20 32 36	−19 25 38
10. 5.84	14 43 47	−13 7 60	10. 4.86	16 31 50	−19 52 25	10. 3.88	18 7 25	−22 18 22	8. 2.90	19 26 23	−21 42 13	**9. 1.92**	20 37 15	−19 9 13
20. 5.84	14 40 56	−12 55 39	20. 4.86	16 30 6	−19 47 40	20. 3.88	18 9 18	−22 17 34	18. 2.90	19 30 52	−21 33 20	19. 1.92	20 42 3	−18 51 50
30. 5.84	14 38 18	−12 44 46	30. 4.86	16 27 49	−19 41 59	30. 3.88	18 10 31	−22 16 50	28. 2.90	19 35 1	−21 24 45	29. 1.92	20 46 56	−18 33 44
9. 6.84	14 36 3	−12 35 57	10. 5.86	16 25 8	−19 35 32	9. 4.88	18 11 1	−22 16 16	10. 3.90	19 38 44	−21 16 44	8. 2.92	20 51 48	−18 15 14
19. 6.84	14 34 16	−12 29 40	20. 5.86	16 22 9	−19 28 40	19. 4.88	18 10 49	−22 15 56	20. 3.90	19 41 58	−21 9 34	18. 2.92	20 56 34	−17 56 40
29. 6.84	14 33 1	−12 26 14	30. 5.86	16 19 3	−19 21 41	29. 4.88	18 9 54	−22 15 52	30. 3.90	19 44 39	−21 3 32	28. 2.92	21 1 12	−17 38 23
9. 7.84	14 32 23	−12 25 52	9. 6.86	16 15 59	−19 14 57	9. 5.88	18 8 21	−22 16 2	9. 4.90	19 46 44	−20 58 53	9, 3.92	21 5 35	−17 20 45
19. 7.84	14 32 22	−12 28 38	19. 6.86	16 13 6	−19 8 52	19. 5.88	18 6 14	−22 16 23	19. 4.90	19 48 11	−20 55 48	19. 3.92	21 9 40	−17 4 9
29. 7.84	14 32 59	−12 34 26	29. 6.86	16 10 32	−19 3 49	29. 5.88	18 3 38	−22 16 53	29. 4.90	19 48 56	−20 54 27	29. 3.92	21 13 23	−16 48 57
8. 8.84	14 34 14	−12 43 11	9. 7.86	16 8 25	−19 0 5	8. 6.88	18 0 41	−22 17 25	9. 5.90	19 49 1	−20 54 55	8. 4.92	21 16 40	−16 35 33
18. 8.84	14 36 4	−12 54 35	19. 7.86	16 6 51	−18 57 57	18. 6.88	17 57 33	−22 17 52	19. 5.90	19 48 25	−20 57 10	18. 4.92	21 19 27	−16 24 16
28. 8.84	14 38 28	−13 8 22	29. 7.86	16 5 53	−18 57 37	28. 6.88	17 54 22	−22 18 30	29. 5.90	19 47 10	−21 1 7	28. 4.92	21 21 42	−16 15 26
7. 9.84	14 41 22	−13 24 12	8. 8.86	16 5 34	−18 59 8	8. 7.88	17 51 18	−22 19 3	8. 6.90	19 45 19	−21 6 33	8. 5.92	21 23 21	−16 9 19
17. 9.84	14 44 44	−13 41 43	18. 8.86	16 5 56	−19 2 32	18. 7.88	17 48 31	−22 19 38	18. 6.90	19 42 58	−21 13 11	18. 5.92	21 24 23	−16 6 8

Glossary

Albedo The ratio of the amount of light reflected by a body to the amount of light incident on it; a measure of the reflecting power of a body. A perfect reflector would have an albedo of 1. The albedos of the planets are as follows: Mercury 0.06, Venus 0.76, Earth 0.39, Mars 0.16, Jupiter 0.43, Saturn 0.61, Uranus 0.35, Neptune 0.35, Pluto 0.5.

Allotropy The property in a chemical element of existing in different forms, with distinct physical properties but capable of forming identical chemical compounds. Ozone, for example, is an allotropic form of oxygen.

Altitude In astronomy, the angular distance of a celestial body from the horizon. In conjunction with a measurement of AZIMUTH, it describes the position of an object in the sky at a given moment.

Aphelion The point or moment of greatest distance from the Sun of an orbiting body such as a planet. The opposite of PERIHELION.

Asteroid One of a large number of rocky objects, smaller than a planet but larger than a METEORITE, in orbit around the Sun. Also known as "minor planets", over 99 percent of the asteroids in the SOLAR SYSTEM lie in a belt situated between the orbits of Mars and Jupiter.

Astronomical unit A unit of distance defined by the mean distance of the Earth from the Sun and equal to 149,597,870 km.

Azimuth The angular distance along the horizon, measured in an eastward direction, between a point due north and the point at which a vertical line through a celestial object meets the horizon. (This is the normal convention for an observer in the northern hemisphere; other conventions are sometimes followed.) *See also* ALTITUDE.

Barycenter The center of gravity of a system of massive bodies; the barycenter of the Earth–Moon system, for example, lies at a point within the Earth's globe.

Black body An idealized body which reflects none of the radiation falling on it. Such a body would be a perfect absorber of radiation, and would emit a SPECTRUM determined solely by its temperature.

Bode's Law A curious numerical relationship between the distances of the various planets from the Sun.The law is often expressed in the form:

$r_n = 0.4 + 0.3 \times 2^n$,

where r_n is the distance of the planet from the Sun and n is $-\infty$ 0, 1, 2, 3 . . . in turn. The resulting values correspond surprisingly closely with the actual distances, but most astronomers consider this to be merely a coincidence.

Celestial equator The circle formed by the projection of the Earth's equator onto the surface of the CELESTIAL SPHERE.

Celestial sphere An imaginary sphere, centered on the Earth, onto whose surface the stars may be considered, for the purposes of positional measurement and calculation, to be fixed.

Chromosphere The layer of the Sun's atmosphere lying above the PHOTOSPHERE and below the CORONA.

Comet A type of heavenly body in orbit around the Sun, with several characteristics that distinguish it from the planets, satellites or asteroids. Comets typically have highly eccentric orbits, and some of them become bright objects in the sky as they approach PERIHELION, sometimes with a distinctive "tail". They are made up of a "nucleus" with a surrounding cloud of dust and gas which forms the "coma".

Conjunction The near or exact alignment of two astronomical bodies in the sky. Also used to describe an alignment between a planet and the Sun as seen from Earth. When the planet passes behind the Sun, the conjunction is called "superior"; in the special case of Mercury or Venus passing between the Sun and the Earth, the conjunction is called "inferior".

Coriolis effect The apparent deflection of a body moving in a rotating coordinate system. For example, a projectile fired northward from the Earth's equator will appear to be deflected to the east, because the point on the equator from which it is fired will be rotating faster than its target to the north. The Coriolis effect plays an important part in determining the directions of wind and ocean currents.

Corona The outermost part of the Sun's atmosphere. It is visible to the naked eye only during a total eclipse of the Sun, when it has the appearance of a halo around the Sun's obscured disc. The corona is the source of the SOLAR WIND.

Cosmic rays Extremely energetic atomic particles, principally protons, travelling through space at speeds approaching the speed of light. A proportion of cosmic rays come from the Sun, while the rest originate somewhere outside the SOLAR SYSTEM, possibly in violent events in the GALAXY.

Culmination The maximum altitude of a celestial body above the horizon.

Declination The angular distance of a celestial body from the CELESTIAL EQUATOR; one of the two celestial coordinates, roughly equivalent to latitude on the Earth, used to represent the position of a celestial object. *See also* RIGHT ASCENSION.

Doppler effect The apparent shift in the frequency of waves that occurs when there is relative motion between the source and the observer. A receding source will appear to emit waves of longer wavelength (or lower frequency) than it would if it were stationary; with an approaching source the effect is reversed, and the wavelength appears to be shorter (higher frequency).

Eclipse The partial or total disappearance of a celestial body either behind a nonluminous body or into its shadow. A solar eclipse, for example, occurs when the Sun is obscured by the Moon's disc, while a lunar eclipse takes place when the Moon passes through the cone of shadow cast by the Earth.

Ecliptic The circle on the CELESTIAL SPHERE defined by the Sun's apparent annual motion against the stellar background. The ecliptic represents the plane in which the Earth orbits the Sun and, because the Earth's rotational axis is tilted, the ecliptic is inclined to the celestial equator at an angle, known as the "obliquity of the ecliptic", which is equal to about $23\frac{1}{2}°$.

Electromagnetic radiation Radiation in the form of waves associated with electric and magnetic disturbances, which may be manifested in a variety of forms, such as light, X-rays and radio waves, depending on the wavelength. The electric and magnetic components are often represented as two waves oscillating in different planes at right angles to one another.

Elongation The angular distance of a planet from the Sun, or of a satellite from its primary planet.

Equation of time The difference between the apparent solar time and the mean time; the value of the equation of time varies throughout the year from about $-14\frac{1}{4}$ min to about $+16\frac{1}{4}$ min.

Exosphere The outermost region of the Earth's atmosphere, beyond the IONOSPHERE.

Faculae Bright patches on the PHOTOSPHERE of the Sun, normally associated with SUNSPOT groups.

First Point of Aries *See* VERNAL EQUINOX.

Flares Sudden brilliant outbursts in the outer part of the Sun's atmosphere, typically lasting only a few minutes. Generally associated with SUNSPOTS, they give rise to a type of COSMIC RAYS.

Fraunhofer lines Dark lines appearing in the spectrum of the Sun, resulting from the absorption of certain wavelengths of light by elements in the outer parts of the Sun's atmosphere.

Galaxy A large system of stars. The term "The Galaxy" refers to the particular galaxy of which the Sun is a member.

Hertzprung-Russell diagram A graph on which is plotted the LUMINOSITY of stars against their temperature or spectral type. The diagram reveals that for a given spectral type, temperature is not randomly distributed. For the most numerous group of stars (the so-called "main-sequence" stars) the higher the temperature, the brighter, in general, is the star; other groupings in the H-R diagram represent stellar types such as Red Giants and White Dwarfs which do not obey this general rule.

Ion An atom that is electrically charged as a result of having lost or gained one or more electrons.

Ionosphere The region of the Earth's atmosphere, extending from approximately 80 km to 500 km above the surface, in which radiation from the Sun ionizes a substantial proportion of air molecules. *See also* ION.

Librations Apparent oscillations of the Moon as a result of which an Earth-based observer can see the surface from a slightly different angle at different times. Over a period of time, a total of about 59 percent of the Moon's surface can be seen from Earth.

Light-year A unit of distance defined by the distance travelled by light *in vacuo* in a year, equal to 9.4607×10^{12} km or 63,240 ASTRONOMICAL UNITS. In astronomy the more commonly used unit for large distances is the PARSEC, which is equal to 3.2616 light-years.

Limb The edge of the visible disc of a celestial body.

Luminosity The total amount of energy emitted by a star per unit of time.

Magnetosphere The region around a planet within which its magnetic field predominates over the magnetic field of the surrounding interplanetary region.

Magnitude A measure of the brightness of a star or other celestial body on a numerical scale which decreases as the brightness increases. The faintest stars visible to the naked eye on a clear night are of magnitude 6; the brightest have a mean magnitude of 1. The "absolute" magnitude of a star is defined as the apparent magnitude it would have if viewed from a standard distance of 10 PARSECS.

Meridian A great circle passing through the poles either of the Earth or of the CELESTIAL SPHERE. In astronomical usage, the term usually refers to the "observer's meridian", which passes through the observer's ZENITH.

Meteor A small particle of interplanetary material that leaves a bright trail across the sky as it burns up on entering the Earth's atmosphere.

Meteorite The remains of a METEOR that reaches the surface.

Meteroid A small lump of solid meteoritic material in space.

Nodes The points at which two great circles on the CELESTIAL SPHERE intersect; in particular, the points at which the orbit of a body, such as a planet or the Moon, crosses the ECLIPTIC.

Occultation The temporary disappearance of one celestial body, usually a star, behind another, usually a planet or moon. A solar eclipse is a particular case of an occultation.

Opposition The position of a planet in its orbit when the Earth lies on a direct line between the planet and the Sun. A planet is best placed for observation when it is at opposition.

Parallax The apparent change in the position of an object due to an actual change in the position of the observer. Measurement of parallax allows the distances of distant objects to be determined.

Parsec A large unit of distance defined as the distance at which a star would have an annual PARALLAX of one second of arc, and equal to 3.0857×10^{13} km, 206,265 ASTRONOMICAL UNITS, or 3.2616 LIGHT-YEARS.

Penumbra The region of partial shadow that is formed around the region of total shadow when the source of illumination is of finite size. *See also* UMBRA. The term is also used to describe the outer part of SUNSPOTS.

Perihelion The point or moment of closest approach to the Sun of an orbiting body such as a planet. The opposite of APHELION.

Perturbations Irregularities in the orbital motion of a body due to the gravitational influence of other orbiting bodies.

Phase angle The angle defined by the position of the Sun, a body, and the Earth, measured at the body.

Photosphere The intensely luminous layer of the Sun that forms its visible surface.

Plage Bright areas of the PHOTOSPHERE associated with active areas on the Sun, caused by the presence of gas considerably hotter than its surroundings.

Planet One of the nine medium-sized bodies (including the Earth) which orbit the Sun; a similar body orbiting any other star. Unlike stars, planets do not emit their own heat or light from thermonuclear reactions in their interiors. The word "planet" is derived from a Greek word meaning "wanderer": the planets are seen to move against the background of fixed stars. An "inferior" planet is one whose orbit lies within that of the Earth, while a "superior" planet moves outside the Earth's orbit.

Polarization A special condition of ELECTROMAGNETIC RADIATION. Radiation (such as light) may be resolved into two components, one electrical, the other magnetic, at right angles to one another. When the radiation is unpolarized, the components vibrate in every direction, but if the radiation is "plane polarized", all the electrical components are arranged in planes parallel to each other, with their associated magnetic components lying at right angles to them. Other types of polarization, such as "circular" and "elliptical", are also possible.

Quadrature The position of the Moon or an outer planet when its ELONGATION is 90°.

Right ascension (R.A.) The angle, measured eastward along the CELESTIAL EQUATOR in units of hours, minutes and seconds, between the VERNAL EQUINOX and the point at which the MERIDIAN through a celestial object intersects the celestial equator. Right ascension is roughly equivalent to longitude on the Earth, and in conjunction with one other coordinate, DECLINATION, specifies the exact position of an object in the sky.

Roche limit The critical distance from the center of a planet within which gravitational forces would be insufficient to prevent a satellite from being broken up by tidal forces. For a satellite with the same density as the parent planet, the Roche limit lies at 2.4 times the radius of the planet.

Saros An interval of 6,583 days (equal to 18 years 11.3 days) after which the Sun, the Moon and the Earth return almost exactly to their previous relative positions. Consequently, the Saros period marks the interval between successive ECLIPSES of similar type and circumstance.

Sidereal period The time taken for a body to complete one orbit, as measured against the background of fixed stars. *See also* SYNODIC PERIOD.

Sidereal time A system of measurement of time based on the Earth's period of rotation, measured against the background of fixed stars. The sidereal day is taken to begin at the moment at which the VERNAL EQUINOX crosses the observer's MERIDIAN.

Solar constant The amount of energy per second that would be received in the form of solar radiation over one square meter of the Earth's surface at the Earth's mean distance from the Sun, if no radiation was absorbed by the atmosphere.

Solar cycle The periodic variation of solar activity, as manifested in the number of SUNSPOTS, the frequency of solar FLARES and various other solar phenomena. The cycle has an average period of about 11 years.

Solar System The system made up of the Sun, the planets (Mercury, Venus, Earth, Mars, Jupiter, Saturn, Uranus, Neptune and Pluto) together with their satellites, the ASTEROIDS, COMETS, METEOROIDS and interplanetary material.

Solar wind An electrically charged stream of atomic particles, mainly protons and electrons, emitted by the Sun.

Solstices The two points on the ecliptic of maximum or minimum DECLINATION; the times at which the Sun reaches these points along its annual path. The summer solstice (corresponding to the maximum declination) falls around 21 June, the winter solstice (minimum declination) around 21 December.

Spectrum The range of wavelengths or frequencies present in a sample of ELECTROMAGNETIC RADIATION. Visible radiation (i.e. light) may be resolved into its component wavelengths by passing it through a prism; white light will be spread out into a band of colors. A glowing gas under low pressure will emit radiation only at certain specific wavelengths, which appear as bright, isolated "emission" lines in its spectrum; similarly, it will only absorb radiation at these same wavelengths. When radiation is absorbed from a "continuous" spectrum, black "absorption" lines appear. *See also* FRAUNHOFER LINES.

Stratosphere The region of the Earth's atmosphere, extending from about 15 km to 50 km above the Earth's surface, between the TROPOSPHERE and the mesophere.

Sunspots Large transient patches on the PHOTOSPHERE of the Sun which appear black in contrast with the surrounding regions. The number of sunspots varies in a periodic way (*see* SOLAR CYCLE).

Synchrotron radiation Radiation emitted by electrons travelling in a strong magnetic field at speeds approaching the speed of light.

Synodic period The interval between successive CONJUNCTIONS or, more generally, between similar configurations of a celestial body, the Sun and the Earth.

Tektites Small glassy objects, found in a few restricted areas of the Earth, whose origin remains a mystery; believed to be associated with METEORITE impacts on Earth.

Terminator The boundary between the dark and the sunlit hemispheres of a planet or satellite.

Troposphere The lowest layer of the Earth's atmosphere, within which temperature decreases with increasing altitude. It extends to a height of about 15 km.

Tropopause The boundary between the TROPOSPHERE and the STRATOSPHERE.

Umbra The dark central region of a shadow. *See also* PENUMBRA.

Vernal equinox The point on the CELESTIAL SPHERE at which the ECLIPTIC crosses the CELESTIAL EQUATOR from south to north (where the direction is defined by the Sun's motion). Also known as the First Point of Aries.

Zeeman effect The splitting of spectral lines (*see* SPECTRUM) when emission or absorption occurs in the presence of a strong magnetic field.

Zenith The point on the CELESTIAL SPHERE directly above the observer.

Zodiac A belt on the CELESTIAL SPHERE extending by about 8° on either side of the ECLIPTIC, marking the region within which the Sun and the planets are always to be found. The zodiac is divided into 12 equal zones which are named after 12 constellations.

Observing Saturn

The rings and satellites of Saturn are beyond the power of binoculars, and a 7 cm refractor is probably the minimum aperture that will give a reasonable view of the rings. With a slightly larger instrument of the kind commonly used by amateurs—a 15 cm reflector, for example—the view will be magnificent when the rings are well displayed, as they will be during most of the 1980s. Serious observation requires a larger aperture, and 20 cm is the lowest limit for a reflector or 8 cm for a refractor. Generally speaking, Saturn bears high powers well, though it is important to stress that a smaller, sharp image is far better than a larger, blurred one.

Drawings

Saturn is an awkward object to draw properly. Using prepared discs is not convenient in the usual way, because the angle of the ring system varies continuously, and it is true to say that a good drawing of Saturn requires considerable artistic skill. The flattening of the disc should never be neglected, and the size of the disc itself should not be much less than 5 cm.

To make a sketch, begin by putting in the disc and ring outlines. Add the main features, such as the Cassini Division (when the rings are sufficiently wide for it to be seen) and the visible belts. Take great care to put in the shadows correctly—both the shadow of the ring on the globe, and of the globe on the ring. Then change to a higher magnification and put in the finer details, paying special attention to any features visible on the disc. With Jupiter, a drawing has to be completed in a quarter of an hour at most because of the wealth of details and the rapid rotation; with Saturn the observer can be more leisurely, because there will be far fewer visible features on the disc, and any which are seen are likely to be rather ill defined.

It is useful to make intensity estimates of the various belts, zones and other features, using a scale from 0 (white) to 10 (black). The equatorial zone and Ring B, the brightest of the rings, will generally be of brightness 1 or 1.5 on this scale. Observations of this kind are valuable, because Saturn does show definite variations.

The disc

As we have noted, Saturn appears much blander than Jupiter, because of the greater amount of high-altitude haze in its atmosphere. The equatorial belts are always visible with an adequate telescope (except when covered up by the rings), but bright spots are rare. They do occur sometimes, as in 1933, and it is important to study them as intensively as possible, taking timings of their transits across the central meridian of the planet.

Satellites

Only Titan is visible with a very small telescope. With a 7 cm refractor, Iapetus, at or near western elongation, and Rhea are easy, and Dione and Tethys may also be glimpsed; larger apertures are needed to show the rest. The main work to be done is with regard to the magnitudes of the satellites.

Transits and shadow transits of the satellites are not easy to observe, except with Titan, but they are interesting to watch when they do occur and can be studied with a telescope of at least 20 cm aperture (preferably at least 30 cm). Mutual satellite phenomena are very rare, and so are occultations of satellites (or, for that matter, stars) by the rings.

Recording observations

Each observation should be accompanied by the following data: name of observer; aperture and type of telescope; magnification; time (GMT); seeing conditions on the Antoniadi scale 1 to 5 (1 being perfect and 5 very poor); and any special features observed.

In Britain, the national society is the British Astronomical Association, which has an energetic Saturn Section. In the United States, similar work is correlated by the Association of Lunar and Planetary Observers.

Bibliography

General

Alexander, A. F. O'D., *The Planet Saturn, A History of Observation, Theory and Discovery* (Dover Publications, 1980)

Articles

Beatty, J. K., "Rendezvous with a Ringed Giant", *Sky & Telescope*, **61**, no. 1, 76–85 (1981)

Ibid., "Voyager at Saturn, Act II", *Sky & Telescope*, **62**, no.5, 430–444 (1981)

Berry, R., "Voyager 1 at Saturn", *Astronomy*, **9**, 6–22 (1981)

Ibid., "Voyager: Science at Saturn", *Astronomy*, **9**, 6–22 (1981)

Gore, R., "Riddles of the Rings", *National Geographic*, **160**, no. 1, 3–31 (1981)

Ingersoll, A. P., "Jupiter and Saturn", *Scientific American*, **245**, no. 6, 66–80 (1981)

Larson, S. and Fountain, J. W., "Saturn's 'New' Satellites: A Perspective", *Sky & Telescope*, **60**, 356–360 (1980)

Owen, T., "Titan", *Scientific American*, **246**, no. 2, 76–85 (1982)

Pollack, J. B. and Cuzzi, J. N., "Rings in the Solar System", *Scientific American*, **245**, no. 5, 78–93 (1981)

Science, special Voyager mission to Saturn issue, **212**, 159–243 (1981)

Science, special Voyager mission to Saturn issue, **215**, 499–594 (1982)

Soderblom, L. A., and Johnson, T. V., "The Moons of Saturn", *Scientific American*, **246**, no. 1, 73–86 (1982)

Index

Figures in Roman type refer to text entries; figures in *italic* refer to illustrations, captions or tables.

T

U

V

W

X

Y

Z

Photographic Credits
p.51 (1) Lunar and Planetary Laboratory, University of Arizona
p.5 (3) Ronald Sheridan
p.10 (1) Courtesy Royal Astronomical Society Library/Ph. Angelo Hornak
p.10 (2) Ann Ronan Picture Library
p.11 (4) Wilhelm-Foerster-Sternwarte, Berlin
p.11 (5) From Annals of Harvard College Observatory, 1850/courtesy Royal Astronomical Society Library/Ph. Angelo Hornak
p.11 (6) Ann Ronan Picture Library
p.13 (3, 4) Planetary Patrol photographs furnished by Lowell Observatory
pp.28–29 (2) NASA
p.34 (top) Lunar and Planetary Laboratory, University of Arizona
p.34 (bottom) NASA/Ames Research Center
p.52 (3) NASA
p.56 (1) The Mansell Collection
p.56 (2) Ann Ronan Picture Library
p.57 (4) Observatoire du Pic du Midi
p.57 (5) Harvard College Observatory
p.57 (6) A. Dollfus/Observatoire du Piç du Midi
p.58 (1A, 1B) Lunar and Planetary Laboratory, University of Arizona
p.61 (3, 4) NASA
p.63; pp.74–75; pp.82–83 Maps: U.S. Geological Survey (Airbrush rendition by Jay L. Inge)
p.65; pp. 66–67;pp.70–71 Maps: U.S. Geological Survey (Airbrush rendition by Patricia M. Bridges)
p.76 (2) NASA
p.77 (3A, 3B) NASA
p.85 (2) NASA
Jet propulsion Laboratory/NASA: p.17 (3A, 3B); p.24 (1); p.26 (1, 2, 3); p.27 (4); p.28 (1A); p.31 (2, 3, 4); p.35 (bottom); pp.36–37; p.38; p.39 (top left, top right); pp.40–41 (top); p.42 (top right); p.44; p.45; p.47; p.48; pp.50–51; p.53 (4, 6); pp.60–61 (1); p.62 (2); p.63 (3A, 3B); p.64 (2, 3A, 3B); p.68 (2); p.69 (3, 4, 5); p.72 (2); p.73 (3, 4, 5, 6); p.77 (4); p.78 (1); pp.84–85 (3); p.86.
NASA/Space Graphics: p.32 (l); p.33; p.35 (top); p.39 bottom left & bottom right); p.40 (bottom left); p.42 (top left & bottom); p.43; p.46; p.52 (1, 2).

Illustrators
Kai Choi, Chris Forsey, Wolfgang Mezger, Colin Salmon, Charlotte Styles